PROCEEDINGS OF THE THIRD CONFERENCE ON

ORIGINS OF LIFE

PLANETARY ASTRONOMY

PROCEEDINGS OF THE THIRD CONFERENCE ON

ORIGINS OF LIFE

PLANETARY ASTRONOMY

Edited by

LYNN MARGULIS

SPRINGER-VERLAG NEW YORK • HEIDELBERG • BERLIN
1973

Origins of Life, Volume III

This Conference was organized and conducted by The Interdisciplinary Communications Program of the Smithsonian Institution and was supported by the National Aeronautics and Space Administration under contract number NSR-09-015-044. It was held at Pacific Palisades, California, from February 27-March 1, 1970.

Library of Congress Catalog Card Number: 72-91514

ISBN-13: 978-3-642-87743-8 e-ISBN-13: 978-3-642-87741-4
DOI: 10.1007/978-3-642-87741-4

Softcover reprint of the hardcover 1st edition 1973

CONTENTS

PARTICIPANTS

Dr. Norman H. Horowitz, Chairman; *Biology Division,* California Institute of Technology, Pasadena, California

Dr. Ian Kaplan, *Department of Geology,* University of California, Los Angeles, California

Dr. Harold Klein, *Director of Life Sciences,* Ames Research Center, Moffett Field, Mountain View, California

Professor Conway B. Leovy, *Department of Atmospheric Sciences,* University of Washington, Seattle, Washington

Dr. Lynn Margulis, *Department of Biology,* Boston University, 2 Cummington Street, Boston, Massachusetts

Dr. Stanley L. Miller, *Department of Chemistry,* Revelle College, P. O. Box 109, La Jolla, California

Dr. Bruce Murray, *Division of Geology,* California Institute of Technology, Pasadena, California

Dr. Leslie E. Orgel, The Salk Institute for Biological Studies, P. O. Box 1890, San Diego, California

Dr. Juan Oró, *Department of Chemistry,* University of Houston, Houston, Texas

Professor Tobias C. Owen, *Division of Earth and Space Science,* State University of New York, Stony Brook, L.I., New York

Dr. Alexander Rich, *Department of Biology,* Massachusetts Institute of Technology, Cambridge, Massachusetts

Professor J. William Schopf, *Department of Geology,* University of California, Los Angeles, California

Dr. M. C. Shelesnyak, *Director, Interdisciplinary Communications Program,* 1025 15th Street, N. W., Washington, D. C.

Dr. Eugene M. Shoemaker, *Division of Geology;* California Institute of Technology, Pasadena, California

Dr. S. Fred Singer, *Deputy Assistant Secretary for Scientific Affairs,* U. S. Department of Interior, Washington, D. C.

Dr. Laurence A. Soderblom, *Division of Geology,* California Institute of Technology, Pasadena, California, and U. S. Geological Survey and Center for Astrogeology, Flagstaff, Arizona

Dr. R. J. Strominger, *Biological Laboratory,* Harvard University, Cambridge, Massachusetts

Professor Gerald Wasserburg, *Division of Geology,* California Institute of Technology, Pasadena, California

Dr. Richard S. Young, *Chief, Exobiology, Bioscience Programs,* Office of Space Science and Applications, NASA Headquarters, 400 Maryland Avenue, S. W., Washington, D. C.

FOREWORD

This is an edited record of the dialogue of eminent scientists attending the third conference in the series on the Origins of Life, supported by a grant from the Biosciences Program of the National Aeronautics and Space Administration. The first conference at Princeton, 1967, was held under the direction of Dr. Frank Fremont-Smith at the time when the Interdisciplinary Communications Program (ICP) was associated with the New York Academy of Sciences.

In 1968, ICP was integrated into the Office of the Assistant Secretary for Science of the Smithsonian Institution, and the entire operation was set up in Washington, D.C. The second conference, also in Princeton, was held in 1968. (See Margulis, ed. 1970 and 1971 for previously published proceedings.) The third conference was held at Santa Ynez, California, Feb. 27 – March 1, 1970.

The proceedings are recorded and edited by the Interdisciplinary Communications Associates, Inc. (ICA, a nonprofit foundation), for ICP. Dr. Lynn Margulis, the Scientific Editor of the series, has been assisted by Barbara Miranda. Harriet Eklund is the ICP Staff Editor.

ICA was formed to encourage effective interchange and interaction among the various scientific and social disciplines and to aid in the solutions of scientific and social problems. Currently its primary concern is with assisting ICP.

M. C. Shelesnyak, Ph.D.
Director, Interdisciplinary
Communications Program
Smithsonian Institution
Washington, D.C.

EDITOR'S PREFACE

This Third Conference on the *Origins of Life* inaugurated a new editorial procedure. Previously the 400 page verbatim transcript had been sent, unedited, to all the participants for their modifications. All the individually edited materials collected by the ICP staff were then returned, collated, and sent to the Scientific Editor for synthesis, content verification, and reference check. At this stage, the manuscript was edited by the ICP Staff Editor before being sent to the publisher. This year, however, at the suggestion of Dr. Shelesnyak, Director of the ICP, and others in the program, we agreed upon a more efficient and effective alternative. Before any participants saw the transcript, I cut extensively, modified and clarified the verbatim text. In the meantime, the references and figures were solicited from each conferee and concurrently each was given the option of editing the *revised* text within a limited time period. Some conferees declined this option. Therefore, although the text has been made more concise and clear and many of the redundancies inherent in oral speech removed, I may have unwittingly introduced errors. The precision and care which generally characterizes scientific writing can only be approached in conversation among scientists – this has been my goal. For documentation or clarification of any issue, the primary scientific literature of the references or the individual participants should be consulted. It is hoped that the reader will accept and enjoy this publication in the spirit it was intended: as an informal dialogue among scientists on the implications of lunar and planetary studies in determining the origins of life.

Many photographs of the lunar surface and other data not reproduced here were shown at the conference, especially by Drs. Murray, Wasserburg and Shoemaker. The pressures of time and production costs make it impossible to publish all of the slides mentioned during the discussions. The reader is referred to the original references as noted and should assume that all of the figures referred to in this text are available for further study.

I acknowledge with gratitude the expert editorial assistance of Barbara Miranda.

Lynn Margulis, Scientific Editor
Boston University

March 1971

INTRODUCTORY REMARKS

Friday Evening Session

The first session of the Third Conference on Origins of Life, held at the Santa Ynez Inn, Palisades Park, California, convened Friday evening, February 27, 1970. Dr. Norman H. Horowitz presided.

SHELESNYAK: Ladies and gentlemen, I welcome you to the third in our series of dialogues on the Origins of Life. Some members have not yet arrived, but my impression is that we have enough talent to make this a smashing meeting–as my British friends would say.

I would like to greet old friends, and tell new ones about this show. We think it is special. First, we want you to know that this is not an ordinary formal conference. There are no formal presentations and we are reluctant to let anyone speak for too long. We feel that in order to understand interdisciplinary problems, participants should make opportunities to interrupt. If there is something obscure in the dialogue or if there is something to which you object, don't apologize for interrupting.

We want a freewheeling, relaxed discussion that bridges the language and idea barriers. I think this group is somewhat different. It shouldn't have too many of either of them.

SCHOPF: We don't have too many ideas or too many barriers?

SHELESNYAK: Too many barriers. One problem is that there *are* too many ideas.

We were a little concerned that our Chairman, Norm Horowitz, would be unable to act as Chairman. We had a good man in the pinch-hitting box, and in spite of my last letter announcing that Dr. Young had been coerced into taking over, Norm's youth and his tremendous healing capacities have taken over instead. We are happy that Norm can carry the ball.

This first evening we will talk about our relaxed freedom. Although many people here know each other, it has always been a practice to make self-introductions. We want you to tell us how your interest developed in your particular area. Often, even though people have been working side by side, they discover at the opening session something has happened to their colleague about which they were unaware.

This first evening we hope people will give some impression of what they plan to give and what they plan to get out of the discussion.

I would like to ask Norm Horowitz to set the pattern for autobiographical remarks and then we can go around the table clockwise as viewed from above.

HOROWITZ: O.K., Shelly. I am Norman Horowitz and this is my third meeting here. I became interested in the origin of life *via* genetics. Subsequently in the space program I became convinced that the search for life on Mars was worth some effort. For the past 5 years I have been spending half-time at the Jet Propulsion Laboratory organizing a Bioscience Section and developing instruments to fly to Mars in search of life.

At the outset of this 5 year series we agreed to alternate these meetings between pre-biological chemistry or evolutionary biochemistry on one hand and space exploration on the other. This year is one that is chosen for space data because 1969 was an extremely important year for acquiring new information.

SHELESNYAK: The year of the Dog.

HOROWITZ: Yes, 1969 may turn out—at least in terms of public interest and support—to be a high watermark for the space program.

Now that dust has settled down, I think it is important to examine the data to find out what the data tell us about biological problems, the future space program, and specifically what we might do in the '71 Mars opportunity and the Mars Lander—which has been delayed until 1975. What are biologists interested in acquiring from future Apollo samples and from beyond Mars, what about the outer planets and Mercury and Venus?

I hope to learn these from the program and also find out what people are doing in their laboratories. I hope to profit by the gossip as well as by the formal gospel.

MARGULIS: I am Lynn Margulis, from Boston University. I also have been here twice before. But considering my genetics background and my lack of qualifications to discuss these problems of planetary astronomy, I think this time I was invited as a gesture of tokenism for the Women's Liberation Movement.

Mrs. Swanson, our stenographer, diligently types every wise and unwise word you say. At the end of the conference—I don't know why there seems to be a conspiracy—I have to make sense out of the transcript, and this is probably the real reason I have been invited again.

HOROWITZ: You are here because we love you.

MARGULIS: Thanks. What can I say after that?

My interests are in microbial evolution and they overlap with those of Bill Schopf and others interested in Precambrian micropaleontology. I know nothing about planetary astronomy and the possibility of life on other worlds, so I am here to learn.

SCHOPF: Presumably there is a lot of Precambrian out there but I don't know how many paleontologists.

KLEIN: I am Chuck Klein, at the Ames Research Center at NASA. Frankly I don't know why I was invited. I am glad to be here–I have a lot to learn–but I've never been here before.

Norm spoke about the manned and unmanned programs, and how and where they are going to meet. I think I have something to learn and maybe even a few things to contribute on that score.

My background is in microbiology, microbial physiology, and I have a very specific interest in exobiology. For reasons I don't quite understand I became involved with Norm Horowitz along with other members of the Biology Team for the Lander on the '75 mission.

I will try to contribute to the meeting as much as possible. I really look forward to learning a lot.

SHELESNYAK: We plan to learn from you too. I think we will see that participation will be two ways. Otherwise it will boil down to nothing.

MARGULIS: Is Chuck Klein the H. P. Klein in the Biology Department at Brandeis University?

KLEIN: Yes. He has changed a lot, though.

LEOVY: I am Conway Leovy, a meteorologist by trade, currently at the University of Washington where I have been for the last 2 years. Before that I was here at Santa Monica at the RAND Corporation where I worked on theoretical models greatly related to circulation of the Mars atmosphere. One thing led to another and since then I have become involved with Norman Horowitz in the Mariner '69 TV experiment, as well as–if it goes–in a surface meteorology experiment for the '75 lander.

I know nothing whatsoever about the origin of life but I'm planning to learn. My interest stems from an interest in present atmospheric processes, their history and their future. I'm particularly interested in the atmosphere of planets–especially the terrestrial planets–because I feel they undoubtedly have a lot to tell us about the earth.

SCHOPF: I am Bill Schopf, in the Department of Geology at UCLA. I have been involved in the Apollo program as a member of the Preliminary Examination Team at Houston to look at the lunar samples when they were first returned. My particular job as a member of a sub-group of the Preliminary Examination Team–called the Bio PET–was to see whether there might be some evidence of biological activity. We looked hard to no evident avail. I have also been involved as Principal Investigator in the Apollo Program. My primary interests lie more in terrestrial biological systems; I'm primarily an evolutionary biologist interested in evolutionary questions.

Much of my work has involved Precambrian sediments—trying to say something about primitive organisms and how they have changed through time—thereby involving me in questions relating to the origin of life. My studies have involved micropaleontological observations with optical, transmission, and scanning electron miscroscopy and some organic geochemistry.

MILLER: I am Stanley Miller in the Department of Chemistry at the University of California, San Diego. My principal interest is in how you start with a primitive reducing atmosphere, make simple organic compounds which come together, make polymers and finally organize into a structure which we call living. I have done a number of experiments in this area over the past years.

Evidence for life on Mars would help us enormously to understand what happened on the primitive earth. Another aspect of my interest concerns the construction of a wet chemistry amino acid analyzer to go on the surface of Mars but apparently not in 1975.

In addition, I am interested in such things as gas hydrates, one of which, the carbon dioxide hydrates, appears to be in the ice cap of Mars. I am also interested in other things but these wouldn't be related to our topic.

ORGEL: My name is Leslie Orgel. I am associated mainly with the Salk Institute at La Jolla but also with the University of California at San Diego. I started life as a theoretical inorganic chemist, more recently have gotten interested in prebiotic chemistry, which is pretty much equivalent to the chemistry of aqueous solutions. Over the past 5 or 6 years I've done experiments in the same general areas described by Stanley Miller.

I don't know anything about planetary astronomy and I suspect one reason I am here is perhaps because once one's name is on the list, it never gets off. Mine got on when we discussed organic chemistry and I guess I am just a fossil left over from previous meetings.

OWEN: I am Toby Owen. Presently I am visiting at Cal Tech. I am a planetary astronomer, although I hesitate to say so since I seem to be the first person to have done so.

I make ground-based observations of planets and comets, trying to characterize the environment as well as possible, specifically in terms of the composition of the atmosphere of the inner and outer planets. Since the development of life is obviously one of the more subtle aspects of the development of a planetary atmosphere, I have to be somewhat concerned with it, although I certainly know much less than I would like to, which is one of the reasons I am here. I hope to contribute some information about the environments from the inorganic point of view.

ORÓ: I have practically fallen asleep. I don't know whether it is the wine or the water in the wine.

It looks like we are all in the wrong place. Everybody is ignorant; maybe we can learn. Presumably if those who are ignorant talk with each other, they have the possibility of learning something more than they already know.

My name is Juan Oró. I am a chemist, a biochemist, a father, and pretty soon I am going to be a grandfather.

OWEN: You are very well qualified.

SHELESNYAK: If you are not a student then you are a practitioner in the origins of life.

ORÓ: My wife keeps telling me that she knows more than I do.

I am at the University of Houston in Houston, Texas. We have been involved in attempting to answer the question of the origin of life on the earth, first as a dialectic exercise and a philosophical problem. Realizing that philosophy was incapable of solving the problem in the twentieth century, we have tried to see whether or not science can make any contribution.

Recently, we analyzed some of the lunar samples to see if they elucidate the matter and with these two colleagues here I'm involved in sending an instrument to Mars in 1975 to determine if any organic chemistry, perhaps any biology, has gone on on that planet.

I must say that results from the moon have not been inspiring for biological chemistry. However, I do think they say something about what happened before prebiological chemistry may have started. The big question in my mind is what we are going to find on Mars. I am not very optimistic. If analogies have any significance–which they don't–I don't think we will find much more there than on the moon. Perhaps this is not the best way to start a symposium among people who are trying to look for life on Mars. But when I see that Mars is saturated with craters I realize that if everything else would allow life, perhaps continuous bombardment may have obliterated it.

Here we may be able to find out whether or not Mars resembles the moon. Perhaps Mars has not been bombarded the way some lunar areas have been and we may find some real organic compounds.

SHOEMAKER: I am Gene Shoemaker from Cal Tech and, unlike the others, I know why I am here. I have been called all sorts of things but one thing I haven't been called is a biologist.

If I understand my charge correctly from Norman Horowitz, it is to bear some sort of witness about the moon and to see what my biological colleagues make of the data. This will keep me in suspense because it is not clear whether much can be said except that we can get some important information about the very early stages of planetary evolution by studying the moon.

STROMINGER: I am Jack Strominger from Harvard. I am here because I opened my mouth once too often and indicated interest in microbiological problems that might be involved here. I am a microbiologist. Most of the organisms I work with are pretty ordinary, but some have been fairly exotic. I am interested in what kinds of experiments are being designed and whether I can contribute to the design of biological experiments.

YOUNG: I am Dick Young. I run the Exobiology Program at NASA Headquarters in Washington. I know exactly why I am here. I am here because I know how life began and I am going to tell you if you are right—at the end of the symposium.

By definition, I have to know. By training I am an experimental embryologist, which is the beginning of life in a sense. By my experience in the last 10 years I have been what we now call an exobiologist which has to do with life under very strange situations. My interest is broadly based because exobiology is a very broad subject. I think we learn as much from the absence of life by studying extraterrestrial situations as we do from the presence of life. The objectives of planetary exploration have comparatively little to do with the presence of life. We don't have to detect life to learn how it may have begun.

I am concerned about what we can learn about the relationship between a planet and its biota in its broadest sense: What happens to a planet as the result of its biota, what happens during the evolution of a planet which either leads to or causes the destruction of its biota? This is a very primitive problem and a very contemporary problem at the same time.

SHELESNYAK: I am Shelly Shelesnyak and I am Director of the Interdisciplinary Communications Program of the Smithsonian Institution. This series of discussions as well as a number of others is sponsored by funds from NASA out of Orr Reynold's office. Orr, unfortunately, at the last minute couldn't join us. However, I think it is more important to tell you a little bit about my own background in science because when I discuss this program and people ask me what sort of subjects we have, I always point out that I do not get involved in any areas or subjects with which I personally have no identification. I have been one of these people whose enemies always accused me of having spread myself too thin and, since I have no friend who has said I am a modern renaissance man, I have just been involved in a large number of things. My original training had been in the biology of reproduction. As Dick's beginning in experimental embryology, mine, too, is a sort of origin of living, if not necessarily origin of life. Back after World War II, Orr Reynolds and I at first, and shortly thereafter with Sid Geller, were all in the Office of Naval Research, even before it officially became the Office of Naval Research. I believe I started the first program for human ecology in the United States Government but, since in 1946 no one knew what in the hell I was talking about, it died on the vine.

My professional involvement ranges from Arctic exploration to child care and development. From 1950 to 1967 I was at the Weizmann Institude in Israel developing an interdisciplinary program in the biology of reproduction. I was fortunate enough to have the Ford Foundation and the Population Council—over a period of about 3 years—give me better than $1.75 million to set up a laboratory to run the program. I was even more fortunate at the end of 6 or 7 years to have adequate staff and a good man to take over so that I could go into a new venture. This led to an invitation by Dr. Reynolds and Dr. Geller to take over this Interdisciplinary Communications Program that was derived from the Macy Foundation under Dr. Fremont-Smith. We changed our sponsoring organization from the New York Academy of Sciences to the Smithsonian Institution. The first meeting I attended as an observer was at Princeton in May '67, where I was introduced by Dr. Fremont-Smith, and they were all happy to have me there. I don't think I put on a very good show because at that time I was listening hourly through a small earphone to news about the Middle East and Israel. I came to the States for the 3 days and flew right back to Israel so I could guard my Institute, which meant spending 7 days and 7 nights in my building. Since I designed it carefully, with a small kitchen facility right next to my office and a shelter in the basement, it was as good a place as any to spend the war.

I am trying to use this program, with modern communication technology, to bridge the communication and idea gap that exists in frontiers in the development of scientific problems. I want to improve communication and use our science and technology toward the solution of social problems. This particular series doesn't quite fit the pattern but we are becoming progressively more concerned with problems related to population, pollution education, health services, and the like.

SCHOPF: How many such programs do you have?

SHELESNYAK: At the moment there are four programs per year in progress, another one is in the final stages of arrangement with the National Cancer Institute, and another one is under discussion with the Office of Education. We can't run more than five or six programs effectively, anyway.

In areas like this, one meeting a year is about maximum. People need an opportunity to develop their own laboratory work or to seek more information. On other problems there is an accumulation of information that hasn't been adequately disseminated in order to have impact. For example, in regard to pollution, we planned four meetings a year over the period of the next three or four years for the Office of Education.

As Lynn Margulis indicated, we take this transcript, send it back to you for revision of your comments, beg you not to revise other people's comments, and then the scientific editor–in this case, Lynn, and maybe she will have to ask for help from some of the planetary physicists or geologists–will work it up into proceedings that will retain the spirit of both the content and the process.

This tradition I inherited from the Fremont-Smith approach, but I am working now on developing other media for dissemination, including certain types of filming, tape recordings, and condensations other than the conversational type.

SCHOPF: Are there any data available on how effective this means of communication really is, the number of books that have been published and distributed for each proceeding?

SHELESNYAK: We are now studying this. It is difficult because one indication–not necessarily a good one– was on the sales of books, but until recently books were distributed free. We have tried to get reactions and there seem to be two audiences that are genuinely interested. Participants find these books are useful a second, third, fourth year later, and graduate students find them interesting and useful.

I think the finest tribute was paid by a critical reviewer of our Fourth Conference on Marine Biology who said that by reading carefully–and sometimes between the lines–he got a great number of new ideas.

Assessing our publication program is extremely difficult. My feeling is that it is nowhere near as valuable as the effort it entails. Yet we do have social-scientific responsibilities to both the community and to the sponsoring agencies.

HOROWITZ: What is the status of Origins of Life I?

SHELESNYAK: It is at the printer's–ready to go to press–and should be out within a short time.

Norm; is there anything more you would like to say?

HOROWITZ: My doctor said I could come to this meeting as long as I don't read and don't engage in actual physical combat.

SHELESNYAK: I look forward to a very illuminating 2 days here. I have never heard so many confessions of ignorance. Someone said that this would give us opportunity for learning, but I am sure you will all prove overly modest.

Your folders contain a reprint of the "1973 Viking Voyage to Mars."

The conference recessed at 9:15 p.m.

THE MOON

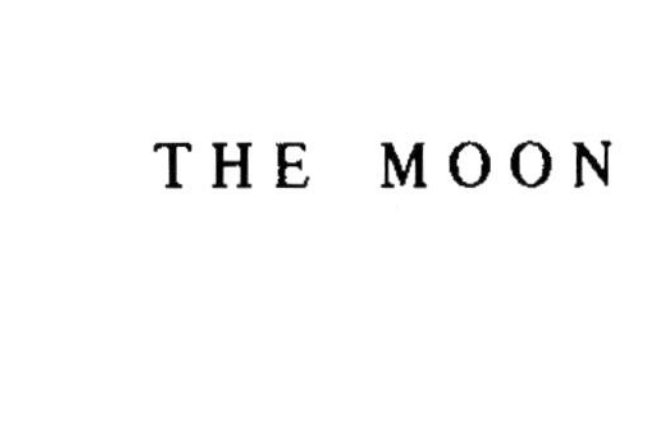

THE MOON
Saturday Morning Session

THE LUNAR SURFACE

The conference was called to order at 9:20 a.m. by Dr. Shelesnyak.

SHELESNYAK: We have four people who weren't here last night, so I'll ask Alex to open this morning's session by what Sir Sollie Zuckerman calls an *ICP Oxford Confessional Modified.*

RICH: To explain why I am here?

SHELESNYAK: Yes, and you can tell us about biological, social, and professional parameters.

RICH: A very complex subject. I was born and educated briefly on the East Coast, moved to the West Coast, spent 5 years at Cal Tech learning about science and related things, and then went back to the East Coast to NIH for 5 years. I have been at MIT now for a little over a decade.

In 1946-47 I became interested in the nucleic acids for not entirely rational reasons. Then nucleic acids were thought to have an important biological function: maintaining the internal viscosity of the cell. Leslie Orgel said it was his thought and I was just concurring for the moment.

It took several years to realize there was more to nucleic acids than simply maintaining viscosity. I was interested in measuring the molecular weight of DNA. As a student in the late forties, early fifties, I observed that the molecular weight of DNA was a time dependent function: Every year papers were published indicating the molecular weight went up a factor of 8 to 10. A very good semi-log plot of DNA's molecular weight with time can be made: It is almost linear and then it suddenly hits the top and plateaus. Then people realize that the DNA molecule is all one piece in bacteria.

My interest in our subject was stimulated by agreeing in a weak moment to write an article for Albert Szent-Gyorgyi, Marine Biological Laboratory, Woods Hole, Mass. I decided to write on the origin of life because all of the explanations I had seen were upside down, that is, they said the first proteins were made abiotically, and then afterward the enzyme catalysts were made. You then make the nucleic acids and the system takes off. This seemed to be absurd enough to warrant a counterstatement.

My interests in both the nucleic acids and the origin of life have remained, bringing me here.

SHELESNYAK: Dr. Wasserburg, we have a tradition of self-introductions as you have just heard. You are cordially invited to make similar or parallel remarks, although you need not necessarily limit yourself to nucleic acid.

WASSERBURG: Saying why I am here?

SHELESNYAK: Who you are and why you are here.

WASSERBURG: You know why I am here–you twisted my arm, that answers that.

As to who I am: My kid just wrote something for school: My father's name is Gerald Joseph Wasserburg. I think that is true. My principal scientific interests mostly concern problems of physics and chemical physics applied to natural systems: earth sciences and planetary sciences occasionally, cosmologic problems and, in particular, time scale problems and evidences of nucleosynthesis.

My interest in the origin of life stems from an unfortunate youthful contact with Stanley Miller when we went to school together and because of his running electric discharges through things. I used to look at them, having fun. Because of regard for Norm Horowitz, I came.

SHELESNYAK: Incidentally, we apparently have made progress. The original concept and title of this program was in the plural form: origins of life, yet both you and Alex mentioned the origin of life. Have we settled that there is one system, or are my ears just too sharply attuned to terminal s's?

I would like to introduce Larry Soderblom, and let him carry on with his own role. He is a graduate student on the verge of completion of his responsibilities in taking his degree in planetary sciences in the Cal Tech Geology Department with Dr. Murray.

SODERBLOM: My principal interests have been in the history of the surfaces of the extraterrestrial planets, terrestrial planets, and the moon in a geologic sense. I spent a summer with Dr. Murray and Dr. Sharp trying to interpret some of the Mariner photography in terms of the history of the Martian surface.

My thesis, which I am presently finishing, is concerned with the evolution of the various lithologies on the lunar surface. It involves two observational techniques. One is designed to recognize and map certain physical qualities of surface materials in order to identify their units of compositional homogeneity. Then from knowledge of the effects of impact on the surface, the studies of overphotography will provide information as to the relative ages of these lithologies.

I hope to pursue this type of investigation further in mapping the history of the moon.

HOROWITZ: Incidentally, Ian Kaplan who will contribute to this program from UCLA is supposed to have been here at 9 a.m. He will presumably arrive. Before Gene Shoemaker takes over, I thought I might say a few words about why biologists ought to be interested in the moon.

There are several reasons that the moon and lunar findings are on the program today. First we know much more about the moon than we know about any other extraterrestrial body. One of the important issues about the moon for biologists is the age of the lunar surface. This is important because it reflects directly on our interpretation of the Martian data which we will discuss tomorrow.

The age of the Martian surface and craters is of biological importance because if we knew the age of the Martian craters we would know how long it has been since there was an ocean on Mars. After the Mariner 4 mission, there was discussion in the literature concerning the implications of the craters on Mars for the age of the surface. There were two schools of thought: The crater frequency on Mars implied a very ancient and possibly primordial surface which never had oceans of terrestrial magnitude which, although pessimistic, would be very meaningful for biology.

The other school of thought was that the age of the Martian surface is much younger than it appears to be from the crater frequency because Mars suffers more meteoritic impacts than the moon. The moon has always been the reference point in these discussions.

We now have some absolute datings from the moon. Dr. Wasserburg from whom we will be expecting to hear, among others, has worked on this problem. I think the outcome of this important problem will very much influence our understanding of Mars.

Another problem about the moon that is of great interest to biologists concerns its origin. I don't know much about this, but I know there have been at least two principal theories about the origin of the moon. One is that it was once a part of the earth and was pulled or centrifuged out of the earth after the formation of the earth. The second theory is that the moon was never part of the earth but was a separate body captured by the earth after both bodies had formed.

RICH: Isn't there a third theory, that they both condensed at the same time but were not together before they condensed?

HOROWITZ: Is that right, Gene?

SHOEMAKER: That was once a very popular theory.

HOROWITZ: Some suggest the moon is kind of a fossil. If it originally was a piece of the earth, in principle the moon could tell us about the primordial earth since on the moon nothing much has happened since its formation. It has no atmosphere and, except for the irradiation by solar radiations of the upper layer, presumably not a great deal of chemistry has occurred there.

If the moon were a chunk of the earth it would be important in telling us something about the early composition of the earth.

It also would be terribly important in understanding the origin of life and the early history of the earth if we could know whether such a cataclysmic event as the removal of one percent of the mass of the earth had ever occurred. If it originally was part of the earth the discovery of some remnants of primitive life on the moon is even possible.

But if the moon was formed someplace else in the solar system or galaxy and it was never part of the earth, this is of great interest. Was the moon formed in the same crucible as the earth; is it made of the same materials? If it is made of the same material as the earth, this gives us clues about the nature of the chemistry that may have occurred on the primitive earth. But the geochemistry and geophysics of the moon may suggest the moon was formed in some other place far away from the region of the solar nebula where the earth was formed. If so, we may be a little more doubtful about drawing conclusions about the primitive earth from the findings on the moon. These questions are of considerable biological interest.

The question of the extent and kind of chemical differentiation on the moon is another of interest to biologists. Current views associate the origin of life on earth with its geological differentiation into core, mantle, and crust. It is generally believed that the process which resulted in differentiation of the earth geologically also produced the oceans and the atmosphere. In this atmosphere of reducing volatiles, the photochemistry that preceded the origin of life presumably began.

This process of differentiation–release of volatiles from the interior of a rocky planet–is of primary importance for understanding the origin of life. The earth has been the only body available for study, but now it is possible to carry out investigations on the moon. We can compare the moon with the earth to get a much deeper understanding of questions such as what relationship the mass of a body has to its differentiation; what the relationship is between chemical composition and differentiation.

Later on in the day we will discuss more obviously biological problems: The results that were obtained from organic analyses of the Apollo 11 samples and various biological experiments that have been carried out on returned-lunar material. Obviously such information can be of unique importance in giving us clues about the origin of life on earth.

About 10 years ago when the space program was just being born, a paper was published by Lederberg and Cowie (1958) which attracted quite a bit of attention proposing the moon was a gravitational trap for granules of cosmic organic matter which these authors suggested might be floating around the solar system. If true, we might find evidence of organic matter of cosmic origin in the returned lunar samples. This information would obviously be invaluable. On the other hand, if analyses show little or no organic matter, then it seems to me this hypothesis would be virtually excluded from further consideration.

RICH: Not necessarily. The rate of accumulation and of degradation and volatization may be such that you would observe nothing in organic analyses even though the moon is acting as a trap.

HOROWITZ: Yes. Maybe we can discuss this; much would depend on the actual age of the surface, and so on.

It has sometimes been suggested that life may have originated on the moon. If the moon originated in the same part of the solar nebula as the earth, it may have had a reducing and aqueous atmosphere during its earlier stages and there may have been the production of organic compounds by the reactions that Stan Miller and Juan Oró and Leslie Orgel and others have been working on. There is even a possibility that remnants of life initially originating there would still be visible on the moon. Some have even thought that there might still be life on the moon. So much so has this thought existed that the whole quarantine program at Houston was based on the assumption that there is a remote possibility that life has evolved on the moon and that that life is actually dangerous to the earth. I have never thought that there was any real chance of life existing on the present moon and I have often said that if life were found in the returned lunar samples I would hand in my biologist's license and get a job on the railroad.

Also, there has always been the question of panspermia. Is there any life on the moon that may have been transported from the earth?

RICH: Is there a difference between astroplankton and sperm plankton?

HOROWITZ: I will lump them together. An astroplankton has been proposed: Life of unknown origin is floating around in the solar system; it falls on the planets and germinates and evolves where conditions are favorable. If true, one might find evidence on the moon. Another proposal has been that meteoritic impacts or volcanic explosions on the earth have ejected terrestrial organisms, giving them sufficient velocity so they have left the earth and reached the moon.

WASSERBURG: That is known as the egocentric concept. Panspermia originated in my backyard.

HOROWITZ: I think those are the main biological issues about the moon. I would now like to call on Gene Shoemaker, Professor of Geology at Cal Tech, Chairman of the Division of Geology, to lead the discussion on the moon today. He is the principal field geologist on Apollo 11, 12, and 13, knows as much and has as good an overview as anyone on the nature of the moon.

SHOEMAKER: Now that you carefully outlined the problems it is clear that my purpose is to give an introduction and you will give all the punch lines.

WASSERBURG: It's pretty hard to talk about these ages without knowing what the surface is like. These subjects are so interwoven that it is difficult to proceed without hearing first what Gene has to say.

SHOEMAKER: I would like to provide background about the lunar surface for some of the details that Gerry and others such as Ian Kaplan, Juan Oró, and Bill Schopf will discuss. In particular I will discuss the nature of the pulverized debris that covers over 99 percent of the moon's surface because almost all of our current evidence on the topics Norm Horowitz has mentioned comes from this. First I'll give an overview; then maybe we can come back and fight about some of the details.

The landing last summer of Apollo 11 occurred in the southern part of this dark Mare Tranquillitatus, that is the Sea of Tranquility–although I prefer these nice old Latin names, it seems they are going to be anglicized.

To one who is not directly connected with study of the moon, it seems that over the past 10 years there has been a lot of controversy about the origin of these maria, the dark plain or seas. However, if you were to have taken a poll and had asked people who have spent any significant amount of time at the telescope, under favorable seeing conditions, what the origin of the maria was, you would have gotten almost 100 percent consistent response. They would agree that the maria are almost certainly formed as a succession of deposits that have been laid down either as liquid lava flows or as fluidized systems of bits of lava and gas–an ash flow. This consistency of opinion is based on things that we knew from work at the telescope for well over 20 or 30 years, and it was beautifully brought out by some of the Orbiter pictures.

A succession of these features that look like flows on the surface are seen locally on the maria. Therefore, it was no surprise to this group–I will call them lunar observers–that some samples returned from this mare surface turned out to be fragments of rock formed by crystallization of a basaltic liquid.

At resolution well beyond that achieved at the telescope, a very consistent texture of the surface of any maria is seen as one gets closer to the moon. This picture [pointing], taken from an Apollo flight just preceding the manned lunar landing, with a 70 mm film and a hand-held camera, shows the distribution of small craters. One crater here is a little less than a kilometer across. There is a distribution of craters smaller than a kilometer in diameter. The surface, as one gets closer and closer to the moon, starts to fill in with resolvable craters. Craters smaller than a kilometer in diameter show a wide range of shape, from very sharply formed features down to a very subtle depression of the surface, just barely detectable in the photographs, with slopes approaching a limited detection. The detection is determined by the contrast in the emulsion and the differences in lighting at a given angle of the sun.

The landing took place just beyond one of these little fresh sharply formed craters, about 200 meters across. Here are the details of the surface going down another order of magnitude in resolution. This picture was taken by overflying of an unmanned Orbiter spacecraft. The astronauts actually had to fly beyond and land here to get out of a large field of very coarse, rocky debris which forms the rim of this crater.

Beyond this debris rim, which extends roughly one crater diameter all the way around, you can see a more normal surface which has a typical population of small craters–right on down to the dimensions of approximately a couple of centimeters.

RICH: What is the scale here, Gene?

SHOEMAKER: This crater is 200 meters in diameter.

MARGULIS: Does it have a name?

SHOEMAKER: Yes, it has a name widely used but not officially sanctioned by the International Astronomical Union. Here are a bunch of names hatched up in the department by Jack Smith–one of the scientist astronauts–just to describe some more prominent features in the target landing areas, so that the astronauts themselves could remember them and look for them as landmarks. This very sharply-formed crater at the west end of the target landing ellipse, used as a landmark, was called West Crater. It will probably be known forever after as West Crater. I guess that is the way names get started.

HOROWITZ: You are sure it is a crater and not a peak?

SHOEMAKER: You have trouble making these craters go down as you look at them. You need to twist your head a little from side to side and tell yourself, "Down, boy," and with concentration you make craters out of them instead of bumps. They really are depressed in the surface.

On this Orbiter picture an important clue as to the character of the surface material is provided by certain craters of anomalous shape. This crater [pointing] was actually visited by Armstrong. These have anomalous shallow flat floors but well developed rims. A study at other sites, particularly around the Surveyor landings, has indicated pretty strongly that the floors are flat because the crater, in the process of its being formed, has bottomed out at a hard stratum or hard layer that underlies a surface layer of much lower cohesion. The depth from the surrounding surface to the flat floor gives the thickness of the layer of weakly cohesive material. The astronauts landed in a little double crater that they could see when they looked out the cockpit of the lunar module.

ORÓ: What is the thickness of these layers?

SHOEMAKER: On the basis of this kind of evidence the thickness ranges from about 3 to 6 meters. Various approaches to models of how degree layers have been formed predict the thickness of the order of 3 meters, just the depth of a crater visited by Armstrong.

Here is the double crater close up—a couple of orders of magnitude better in resolution; we now actually see this layer of pulverized material. It consists of rock fragments ranging in size from about half a meter at this site down to particles of the order of a micron or still smaller. Here scattered over the surface these little bright objects are rocky fragments, typically 5-10 centimeters across. Only about 5 to 10 percent of the surface material is made up of the grains that are coarse enough to be resolved by the naked eye as you stand on the surface. The vast bulk of this material is quite fine grained, with a median grain size down to the range of 30-50 microns.

This double crater has a depth of about a meter and has been excavated entirely in this fine-grain, weakly-cohesive material. It has not penetrated down to the underlying bedrock or down to the wall of the crater. The rim of it is made up of the same kind of stuff seen in the inner crater areas.

This shows the other type of crater, one about 30 meters across, visited by Armstrong. The lunar module is in the background. This is a pile of very coarse rocks. The center was not resolved in the Orbiter pictures; unfortunately the astronauts did not have a chance to get any sample of that particular pile. I interpret that this was derived from the immediate subjacent bedrock material that has been disturbed and broken up, but not as a part of the local bedrock.

YOUNG: Are those rocks or depressions in the wall?

SHOEMAKER: These are craters, again. This is the rim of a crater seen at a glancing angle. This is a crater looking more into it, formed on the wall. Notice even a relatively fresh crater, 30 meters in diameter, has many little craters formed on it. It has a well-defined grazed rim and it is beautiful compared to the majority of craters of this size at this site; they are much more subdued in form and the rim has worn down and is much reduced.

I'll try to give you some feeling for characteristics of this fine grain material. It is loose enough to be disturbed and blown by the exhaust from the descent engine of the lunar module and at the surface here it is streaked with little ridges and grooves. The little ridges are actually windrows of fine particles that lie behind coarser little grains that are embedded in the surface. Also, this gives one note of caution about what might be deduced from a search for water or organic constituents in samples taken close to the lunar module. Of course, a large quantity of combustion products is impinged directly on the surface during the landing. It would not be surprising if some of these combustion products were found in the samples brought back.

SHELESNYAK: Have you been able to compare samples taken under the exhaust with the ones taken a good distance from it?

SHOEMAKER: All the samples were taken relatively close, within about 25 meters, to the landing site of the lunar module, and the best chance of comparing material that is least likely to be contaminated with that most likely to be contaminated would be to compare samples taken with a simple drive tube or core tube driven into the surface where the tube penetrated up to about 2 centimeters from the surface. No samples were taken here [pointing] where you can see this obvious streaking of the ground.

RICH: How far away from the exhaust jet were these ridges visible?

SHOEMAKER: Usually they died out at or a little before the distance of the LEM footpads but the streaks did continue out to the north, the direction from which the lunar module approached the surface, so you can see the track from each extends–left by exhaust as it came in during landing.

RICH: At what elevation was the retrorocket turned off before landing?

SHOEMAKER: In this case, the lunar module had effectively touched down. I have to separate the two missions in my mind.

WASSERBURG: I think they had it on almost to the bitter end.

YOUNG: I think so, too.

SHOEMAKER: In this mission, the engine was not shut off immediately but continued to burn a little after touchdown.

WASSERBURG: The Surveyor camera mirror in the Apollo 12 shot was covered with dust, as were many other parts of the Surveyor. This may have been caused by spraying of dust as the LEM entered, and that was much farther away than the immediate environs.

RICH: The Surveyor dust was from the LEM, then?

SHOEMAKER: May I amend that statement? Jay Reynolds, who worked with me on the Surveyor television experiment, just this past week has been very carefully studying the optical characteristics of some components of the camera. There is a layer of dust on parts of the camera and on other parts of the spacecraft that was laid down as the Surveyor itself originally landed. On that particular landing the little vernier rocket motors were not turned off at the final stages of touchdown. The radar system that was supposed to send a signal to turn the motors off at about 400 meters, lost lock with the surface at about 30 meters above the surface, probably due to specular reflection of the microwave signals from rocks. The engine continued to fire and the spacecraft landed up on the rim of the crater it was bouncing down into, and sprang back up into space and reached an absolene of about 10 meters. Then it landed again three times atop the slope. The camera mirror was open at that time and dust was generally sprayed over part of the spacecraft during each one of these hops. Gradually the engines were turned off by ground command.

WASSERBURG: You conclude the dust was not due to the Surveyor, the main Surveyor rocket was turned off much earlier or much higher up than the equivalent of the LEM?

SHOEMAKER: In the case of the Surveyor II landing, the rockets continued to burn.

WASSERBURG: The main rocket was off.

HOROWITZ: The vernier rockets are the descent propulsion system. The main rocket is jettisoned.

WASSERBURG: Then signals have changed, that is important.

SHOEMAKER: We can test this directly. You can compare the coating of dust and the optical characteristics on the mirror now with those we observed directly with the television camera immediately after the Surveyor had landed. It is clear that the main body of dust on that mirror was due to the landing of the Surveyor.

If you look carefully at the housing of the camera—which was coated with fine particles of dust on the outside as well as on the mirror—it is interesting that there are places on the camera where some of the dust has been swept off. There are little shadows of dust behind bolts and things that protrude from the housing and these shadows point to the landing of the LEM. Instead of coating the camera, the landing of the LEM cleaned the Surveyor off.

WASSERBURG: The signals get changed within three weeks. Holy smoke.

SCHOPF: Why wasn't a major crater formed under the retrorocket on the LEM?

SHOEMAKER: If you make one of several realistic models of the erosion of the surface by the exhaust and then use the model that has approximate identical characteristics of the material, if the LEM came right straight down over one area and was not shifting laterally, you might get a crater up to a few centimeters deep. In this case, the LEM was moving rapidly to the side and the duration of the period of erosion was quite short, so that the depth of any expected crater for this particular landing is considerably less than the inherent relief of the surface, so it would be very hard to recognize such a crater.

RICH: What was the approximate distance from the LEM to the Surveyor on the Apollo 12 landing, where the blast took off some of the dust?

SHOEMAKER: About 125 meters.

KAPLAN: Gene, what is this slide? [pointing]

SHOEMAKER: I think this picture was taken by one of the astronauts during the traverse on the surface, after landing. If you look very closely you see a little crack here in the regolith–I say *regolith* because soil has so many implications. Soil mechanicians wouldn't call this soil. It appears that the explanation for this little break is that the gas from the rocket engine soaked into the ground and then when the engine was turned off it popped back up again, a certain exhaustion of gas that has soaked into the regolith material pops back.

HOROWITZ: Gene, what is the root of the word regolith, what is "rego"?

SHOEMAKER: A mantle or covering: Regolith is just a covering layer of rocky material.

COMPOSITION OF LUNAR ROCKS

This figure shows some coarser rock particles. From the astronaut's rather wide boot–about 10 centimeters–you can get some feeling for the size distribution of particles. Some things that look like grains are actually little soft aggregates of particles. The astronauts' footprints were typically 1 to 2 centimeters in maximum depth. The material is very fine. As the smooth surface was depressed a very smooth imprint was left. This led to considerable confusion. When we first saw this effect on the Surveyor landings it was argued that all the rock fragments must be on the top surface and everything else is fine grain, which is nonsense of course. In fact the coarser particles, which are smaller in volume relative to the finer ones, are being compressed and molded within a finer surrounding material. As you go to depth, a size distribution of particles much like that at the surface is found.

This fine material consists of about 50 percent small rock fragments and individual mineral grains derived from the crystalline rock fragments and about 50 percent glass. Most glass is shock-vitrified mineral grains: crystalline material transformed to glass in a solid state by shock. However a small proportion, maybe 10 to 20 percent, consists of materials that have actually been heated to the melting point. The raising of the temperature has almost certainly been caused by shock, and a significant part of this melted fraction consists of beautiful little glass spheres or broken spheres. Perhaps Gerry Wasserburg will show you more pictures of these. A great many of them are actually bubbles, hollow inside, and they show a wide range of chemical composition.

MARGULIS: What is the diameter of these spheres?

SHOEMAKER: This particular slide shows objects of a few tens of microns.

MILLER: Do you mean millimeters?

WASSERBURG: Those are about 100 microns.

MARGULIS: Is that a photomicrograph?

WASSERBURG: Yes. Glass balls may range from almost half a centimeter down to sub-micron size.

SHOEMAKER: It is hard work to find many as large as half a millimeter across. The bigger ones–some were actually photographed on the lunar surface on the Apollo 12 mission–are, as Gerry mentioned, a couple of centimeters. They are not solid balls but are hollow, they are bubbles of glass.

HOROWITZ: Are those diagonal lines scratches, or what?

SHOEMAKER: This is a ground metal surface, not lunar. Typically, the majority of these little glass balls are 10 to 20 microns; you have to work hard to find them larger than that. With a lot of hand selecting you get some that are hundreds of microns.

LEOVY: Is the idea that these are formed by shock vitrification based on stimulating things in the laboratory?

SHOEMAKER: I was not referring to these hollow spheres when I said shock vitrification. The shock-vitrified material has not been melted, it has simply been transformed to an amorphous form in solid state at temperatures well below the melting point, by shock. This can be done in the laboratory. The grains of glass formed in this way preserve all the original textural features of the crystalline mineral grains. The optical property of the density of this material has been changed and the crystalline grains simply become disordered at scales. They are amorphous both optically and to x-ray, but there is a transition between the crystalline and amorphous material.

There are not just little spheres, but dumbbell shapes, and so on, shapes formed by surface tension of a liquid. Since it has been melted, it is quite different from the bulk of the glass.

SODERBLOM: What percentage of the material has actually been melted into spherules?

SHOEMAKER: Five percent is a good figure. What would you say, Gerry?

WASSERBURG: I think so. There is a problem. The glass balls are noticed first but the glass is in broken bits and chips and all sorts of kooky shapes. You tend to focus on the simple geometrical shapes as glazed coatings.

SHOEMAKER: If you slice these open in some you will find scattered bits of meteoritic material, that is, very tiny spheres of nickel iron: Often included are drops of troilite and schreibersite. This is very direct evidence that these were formed by meteoritic impact.

YOUNG: Is there any gas trapped in them?

SHOEMAKER: Yes, gas is trapped inside. I think Juan Oró will talk about that.

MARGULIS: Any metallic iron?

SHOEMAKER: There is nickeliferous metallic iron, very tiny drops within these spheres. There is also free iron in the rocks themselves, of different chemistry. It does not have the high nickel content to distinguish the meteoritic iron from the indigenous free iron.

KAPLAN: Do you or Juan have evidence of their containing gas? This is the first time I have heard that these spheres contain gas.

ORÓ: Deciduous cavities have been broken out, mainly nitrogen and some amounts of carbon monoxide are produced.

SHOEMAKER: Let's look at some recovered fragments and I'll give you a quick introduction to the rock types that can be identified on the lunar surface.

This rock is only a few centimeters across. It was about half buried in the finer grain material and it was collected in the contingency sample. This short pointed end was sticking up out of the surface, the rounded part of this rock was the part that was buried. It is just the opposite of what you might have guessed if you didn't know the evidence from the photographs. This little light band across here [pointing] turned out to be right at the point where the fine grain material lapped up against the side of the rock. I don't know why it is light, but it looks like some sort of bathtub ring.

HOROWITZ: Was the upper part exposed to solar radiation?

SHOEMAKER: The light band above was the part sticking out of the surface, below was the part buried in the surface.

SCHOPF: How many rocks have been photographed on the lunar surface that can then be identified back at the lab?

SHOEMAKER: From Apollo 11 I am going to show you all three of them. I am hoping to work on that problem a little harder later this year. There is a good chance of finding more.

WASSERBURG: I want to emphasize that this rock is important. If I go blabbing about ages and implications this is the one the age measurement was made on.

SHOEMAKER: This shows a representative of one of the two main rock types. Objects called rocks are simply hard enough to pick up and bang on; they don't immediately fall apart–a rock is just some coherent object made from minerals and glass. The vesicles that can be seen in this rock were formed as gas bubbles in the melt from which it crystallized. These crystalline rocks show a considerable range of grain size and an abundance of vesicles. Sometimes the vesicles are strung out in bands; they merge partly together, giving a pretty good flow banded structure quite familiar to those who study crystalline terrestrial rocks. The gross texture of this lunar rock could be matched very closely by many rocks from the basaltic lava flows in Washington State.

Here is a close-up of the same type with the rock broken open; It has a fresh fracture surface showing the inside of some of these gas bubbles or vesicles. The walls are often beautifully shiny and the specular surfaces are formed by shiny mineral faces.

A thin slice of this rock examined under a petrographic microscope shows vesicles filled in by a cementing material that is dark under polarized light. The wall of the glass bubble can be seen. There are three major mineral phases. The stuff which appears black in the slide is mostly, but not entirely, an opaque oxide of iron and titanium that is called ilmenite. The grains that are white or light gray in this slide are calcium feldspar, and the grains that in thin sections show an array of colors ranging from red to blue—just polarization, they are actually brown in plain light—are pyroxene. These are the three main types; calcium aluminum silicate, the feldspar; a magnesium iron silicate of pyroxene, and the iron titanium oxide, $Fe\ Ti\ O_3$.

There are many minor minerals, too, that Gerry may want to talk about.

The second rock type brought back is not a crystalline rock.

STROMINGER: Did you say all the mineral types in this first rock are similar to those found in rocks in Washington State?

SHOEMAKER: Yes, but they differ in minor details of chemistry. They overlap in composition the minerals in similar surface terrestrial rocks, but the total composition of lunar rock is unusually rich in iron and particularly in titanium, about three times as rich in titanium as similar rocks on earth.

YOUNG: I thought Klaus Keil found a new mineral or nonterrestrial mineral.

SHOEMAKER: Yes, there are some peculiar new minerals, as minor constituents of these rocks. These new minerals reflect a somewhat different environment and chemistry in particular—the low oxygen gases of the melt itself which gives rise, for example, to free iron. There is essentially no ferric iron in these rocks. The free iron is ferrous iron and silicates. One new mineral is a pyroxinoid, similar to but different in chemical detail from the pyroxene I showed you in the earlier slide.

RICH: Gene, what cation in terrestrial rocks has been replaced for the high titanium in the lunar rock?

SHOEMAKER: Titanium hasn't replaced a cation in the silicates, rather it is going into the oxide. We find the same mineral, ilmenite, in terrestrial rock. It is just about three to four times as abundant as the most ilmenite-rich-rocks.

RICH: So it is the same material?

SHOEMAKER: The same material but a different proportion of the three main minerals. Whereas a basalt on the earth would not contain more than about 5 percent ilmenite, in the lunar rocks it is typically about 20 percent.

WASSERBURG: I would like to amplify this because there is something important here. The new minerals are trivial issues, absolutely unimportant, as Gene pointed out, except not so bluntly. The importance as Gene said, is the reflection of a highly reduced condition.

The titanium is present as $FeTiO_3$, which accounts for the bulk of the excess of the titanium. This is sort of mundane. It is interesting that a non-trivial portion of titanium is substituting for Fe or what would be Fe+++, except in the silicates.

HOROWITZ: What is the evidence for the reducing nature of the rock?

WASSERBURG: A variety of things.

SHOEMAKER: The basalt, the ferrous iron, for one.

WASSERBURG: There is no Fe+++.

HOROWITZ: No ferric iron at all?

SHOEMAKER: No ferric iron.

HOROWITZ: Are the terrestrial basalts a mixture?

SHOEMAKER: Yes.

KAPLAN: In terrestrial basalts, especially in those found near the surface, there is atmospheric oxygen and oxygen seeps in.

WASSERBURG: The net result, regardless of why, is that all the lunar rocks show the absence of oxygen.

ORGEL: Is the titanium all Ti^{++++}?

KAPLAN: Yes, so far as we know. Those experiments, which have to be pursued more, are pertinent to the reducing problem. Historically, geologically, and experimentally, there is inordinate ignorance on titanium and silicate systems.

ORÓ: Is it pertinent to point out the measurements made by A. Turkevich–University of Chicago–on the lower titanium content of some other areas?

SHOEMAKER: Yes. The titanium content observed at Tranquility Base is similar to that estimated by Turkevich for the landing site of Surveyor V, about 25 kilometers away. This extremely high titanium content is not necessarily a moonwide feature. All the other surface analyses of the two other localities were simply analyses–one at the Highlands and one at another mare site–and show lower titanium abundance. At Apollo 12, the titanium abundance is also lower than at Mare Tranquillitatis, so the extremely high titanium may be a local anomaly associated with the lavas at the Mare Tranquillitatis.

The titanium abundance at all localities and particularly at the Apollo 12 landing site is still above the range for normal terrestrial basalts. The rocks returned are enriched in certain refractory elements; titanium is the most abundant one, chromium is another.

YOUNG: Can you conclude that the moon, therefore, could not have come from the earth at a certain period of time?

SHOEMAKER: You might, but you might also conclude that materials ultimately condensed into the moon were at one stage heated to the point of selective loss of volatile and moderately volatile elements, and possible residual enrichment of refractories.

HOROWITZ: Such as titanium?

SHOEMAKER: Yes.

RICH: So, the problem is a sampling problem?

SHOEMAKER: Yes, in part. I also think it is too early to draw sweeping generalizations from two samples.

RICH: Since surface material has been worked about, isn't subsurface material more pertinent to the question of any relationship to terrestrial rocks?

WASSERBURG: The first question is whether or not the whole bloody moon hasn't been "worked about." There is the problem about whether the moon is differentiated. If the moon is differentiated in an anterior sense, then it is worked about in a very coarse grain of maybe about a 500 or 800 kilometer scale sense–in which case you get to look at very different things.

SHOEMAKER: Here is another rock about 10 centimeters across that has been identified on the surface. This is a picture of it in the laboratory, a space partly coated with dust.

You may as well get used to the name hung on the other main rock type: microbreccia. The microbreccias consist of fragments of crystalline rocks. The size distribution of this material--from a few centimeters across on down–of the rock particles is just like the size distribution of the rock particles in loose soil. Indeed, as one looks at the chemistry of the microbreccias, at the noble solar wind gasses impregnating the finer constituents of this material, one can go all the way down the list.

The constituents of this rock are the same as the loose particles in the soil. There cannot be any serious doubt that these microbreccias are simply soil that have been made back into a rock again, probably by shock compression. The material has been squeezed, not very hard in most cases, and indurated or reformed into a rock.

ORGEL: Can this be done in the laboratory?

SHOEMAKER: Yes. We are quite familiar with this in cratering experiments and for 10 years we have been calling them *instant* rocks.

MARGULIS: From the quantity or the ratio of ferrous to ferric iron in terrestrial basalt, can anything agreed upon be deduced about the oxidizing state of the atmosphere?

WASSERBURG: Yes, it is a function of the atmosphere. A very significant amount of the different oxidation states of iron, quite typical of the circumstances in terrestrial rocks, must be seen. A very strong statement: that this circumstance in the lunar rocks is a very rare thing, can be made. Before lunar analyses were made, if somebody announced this chemistry—forgetting titanium anomalies—then you would say he had a kooky rock.

MARGULIS: Because of its relative reduction to terrestrial basalt?

WASSERBURG: Yes, extremely highly reduced.

MARGULIS: Are rocks on the earth seen in some systematic way to be less oxidized as a function of age?

WASSERBURG: No. The problem is that the earth has an atmosphere. What is the most surprising thing about lunar rock? Gene showed the clinker. Lunar rock is inordinately fresh. I have never seen a rock as fresh except for a very young rock, like zero age.

HOROWITZ: There is geological evidence for reducing conditions on the primitive earth. I don't know whether it is in basalts but there are uraninites, pyrites, and so on.

MARGULIS: Are ancient terrestrial basalts more like the lunar rocks?

WASSERBURG: No, they are all rotted up. A typical feldspar crystal on the earth is cloudy, opaque, because it has been hydrated and has formed clay minerals. The oxidation state of older rocks is completely altered by their exposure. These rocks are older than any rock on the earth, and yet they look like a zero age rock because they never saw any water.

KAPLAN: The original question was, by looking at a basalt crystallized out in the atmosphere, can we tell the partial pressures of the oxygen and then extrapolate back?

WASSERBURG: In time?

KAPLAN: The question of time is that for the time of the earth the record has been destroyed because of ground water, and atmosphere, and the mess here that we don't have on the moon.

WASSERBURG: Take glass. There are Roman table glasses, characteristically beautiful because they have been vitrified. There are no terrestrial glasses. Some of these lunar glasses are almost as old as the solar system. They never saw water and if you put them in water they would devitrify.

KAPLAN: Did you want to leave the discussion on the origin of the breccia? This is an important question and I noticed you glossed over the instant rock quickly.

SHOEMAKER: Do you want to fight about it now?

KAPLAN: I am not sure the evidence is convincing that those breccias were right on the spot at the place of collection.

SHOEMAKER: From the isotopic studies comes important evidence on this, and I will add a little to the argument.

The class of crystalline rocks of wide variation in grain size can be seen in the microbreccias. Not all rock fragments in the microbreccias or in the loose fine material of the regolith are identical with the coarse crystalline rock fragments. A few percent of the rocks of the class that you will find consist of rock types not recognized in the coarse pieces. It may be that this is in part what Lynn is referring to.

Some of these strangers in the microbreccias and in the loose fines are feldspar-rich rocks constituting a few percent of the total, shown here in polarized light. The dark part is not opaque minerals but isotopic material, giving you a feeling for the abundance of glass. In some cases the glass is so clouded with minute spherules that it is nearly opaque so that the abundance of black on this slide is somewhat greater than the actual abundance of glass.

SHELESNYAK: How do you define the word *glass?*

SHOEMAKER: I am referring to silicate material which is amorphous, both optically and to x-ray. This will lead us to the question implied by Dr. Kaplan. I want to discuss some of the rocks and their individual histories on the surface of the moon. The top of the rock we just looked at is rounded off and densely covered with minute pits. In this particular fragment you can see only the larger ones. The rock is about 10 centimeters overall length, so the pits are from 0.5-4 millimeters across.

A majority of these pits are lined with glass.

RICH: Don't they look like melted areas that Tom Gold—at Cornell University—saw with his fine camera?

SHOEMAKER: Yes, I think so. Not in all cases because some of the melt is distributed not in pits but over patches on the surface. I may have one showing a patch of melt on the surface and the melt could be draped over other things. The melt has been sprayed around but most of the melt on these rock surfaces is in the pits.

RICH: Are they likely to be of similar origin?

SHOEMAKER: Yes.

RICH: Gold was talking mostly about melts, not in rocks but in just the amorphous little craters.

SHOEMAKER: In craters of a dimension of half a meter.

RICH: Do you find those on all sides of the same rock?

SHOEMAKER: Yes.

RICH: Then it makes his explanation less likely.

SHOEMAKER: It surely does.

WASSERBURG: Even before it is published.

RICH: Let me explain. Tommy Gold saw these similar craters lined with what looked like molten material and postulated that this may have been caused by a solar flare or comet, something with a very transient outpouring of energy, massive enough to singe and melt surface features of the moon. I think if what Gold observed was the same as on these rocks, portions of which were clearly pointed away from the sun and buried, then his explanation isn't likely.

SHOEMAKER: I think Gold is the only man alive who believes that theory and has also studied the glass.

RICH: I don't actually think he believes it.

SHOEMAKER: How many of his theories he believes is an open philosophical question.

ORÓ: I think Gold's idea is unlikely but the fact that you find this material all around the rock need not preclude a solar event that happened a repeated number of times.

WASSERBURG: You are not going to melt rocks with solar flares.

I would like to reformulate this; and I don't like the Gold proposition. There are different kinds of glass coverings of glassy materials. Some of them Gene has very clearly shown you and he could show you more. Others are broader coatings which will flow and be gooey like a runny ice cream cone. We don't know the origins of the different kinds of glass and one problem is to see if there is a very deep difference between them. I think to discuss this in terms of Gold's ideas obscures the issues at hand.

SHOEMAKER: I wish I had a picture of one most spectacular rock that was returned from Apollo 12 which is just as Gerry described. It looks as though chocolate had been poured on one side and dripped down on three of the four sides of essentially a tetrahedral-shaped rock. I have a private suspicion how it was formed. I think melt was simply sprayed on that rock. I could show you melt like that from meteor craters. I also think it is shock-melted material, but there is a wide variety of glasses in many forms.

Here is the bottom side of that rock—we saw the top before. It is heavily coated with fines adhering to the bottom, but here are pits. They, too, are glass lined, implying the bottom was once exposed to the pitting mechanism. We had interesting debates, one of which I lost, when these were first described to us by those who were looking at them optically. I think nearly everyone agrees–I certainly do–that they are little high-speed impact craters and the glass in these pits is formed by shock melting. That glass in newly examined cases probably shows contamination, but in some instances it is quite different in composition from the rock itself.

LUNAR CRATERS AND THE REGOLITH

SHOEMAKER: I want to talk about a model of the surface. Let me remind you of the size distribution of the craters. The craters in the foreground are a few tenths of centimeters across and go on down to the smallest in this photograph of the surface–the size of a couple of centimeters, craters on craters right on down to the smallest sizes.

The size-frequency distribution of these craters follows a form already quite familiar to us from Ranger, Orbiter, and Surveyor data. Where we have abundant data and good statistics, craters smaller than about 100 meters tend to follow a simple power function. The thin line is that function originally derived not from this data but from Ranger photographs. The scatter is large simply because they don't have a number of counts. Soderblom may discuss this distribution. For about 5 years we have been referring to it as a steady state distribution of craters. At the Surveyor and Apollo landing sites where we have looked closeup, independent of the abundance of large craters, the size distribution of small craters is essentially identical. It is a time variant of steady state distribution up to some point.

ORGEL: Is it always clear whether something is a crater or not? Do they range imperceptibly from things which are well defined craters to things that you are not sure about?

SHOEMAKER: I use the term craters in a very simple empirical sense as a depression in some kind of surface without prejudice to how it is formed.

ORGEL: Any kind of depression?

SHOEMAKER: Yes.

RICH: There are depressions where the angle of incident light makes them visible and then there are other depressions which would not be visible.

ORGEL: This is the sort of thing I mean.

RICH: The measurements are made when you see the depression. If you don't see it, you don't measure it and it depends upon the angle of the incident light. The data is biased in favor of eliminating older craters with smaller depressions, but that may not affect the distribution because all craters age and become shallow.

SHOEMAKER: Let me amplify this. Including things of extremely irregular shape as possible craters, some surface features obviously have been formed by merging of objects with a roughly circular ground plane. With the very tiny objects distinguishing depressions and the low area between bumps, it is very difficult because the regolith is structured. It is formed by little clods and lumps on the surface. Therefore to talk about "craters" loses meaning if the things are smaller than a couple of centimeters.

HOROWITZ: Do they have to have roughly circular outlines to be called craters?

SHOEMAKER: Yes.

HOROWITZ: Do you call a rill a crater?

SHOEMAKER: No, we are talking about roughly circular objects.

ORGEL: Will two or three people looking at the same picture without any previous collusion agree on the standards?

SHOEMAKER: That has been done many times and the answer is *yes.* A technician at Ames Research Center counted the craters, with no particular hobbyhorses to ride, and came out with the same thing.

KAPLAN: Gene, I recall you constructed a curve like that about 8 to 10 years ago. How has it changed in character and shape with recent satellite observations?

SHOEMAKER: Ten years ago I had to stop with the smaller objects, at the limit of observation of the telescope. Most of this curve has been dependent on the higher resolution data from Ranger, Orbiter, and finally from Orbiter and Surveyor pictures. The only thing I have to back off on is that I am not so sure that this part of the distribution [pointing] is caused by secondary craters.

RICH: How does the relationship of crater height with crater diameter depend on diameter?

SHOEMAKER: That's a good point. As you go to the larger craters, they typically give a very constant crater shape. From about 0.5 or a bit bigger on up to about 10 kilometers diameter 90 percent of the craters have a very standard shape. Craters of diameters less than half a kilometer give you a range of shapes, a softening of the crater forms, and in smaller ones there is more variation in crater diameter to crater depth until finally at this point these two curves fit. I fit two simple power constants to this table and when you get to this region you finally see the complete range of depth to diameter ranges; from this point on down you see at any given size a complete range of size to depth.

RICH: Is it possible that the bulge reflects the phenomenon Leslie mentioned: The shallower craters are not measurable–skewing of the distribution is simply based on the fortuitous fact of the angle of incident light? Do you follow me?

SHOEMAKER: My interpretation is not too different. The difference between these two curves represents to me the number of craters that have been formed and then have been subsequently so modified that they are no longer recognizable. You really expect that the majority of those actually have disappeared and if this curve is extended down to about 10 meters you will run out of surface areas. Just integrating the area of the craters, they have to be formed one on top of the other.

WASSERBURG: Is this a cumulative curve so all the craters are greater than that diameter?

SHOEMAKER: That is correct.

RICH: Has there been enough material accumulated from the Martian pictures to give us analyses of small craters that are comparable to this?

SHOEMAKER: The Martian pictures take you right down into the upper end of this distribution. If you plot the distribution from largest to smallest on Mars, it pales off and does not follow this curve at all for the small craters.

SODERBLOM: The limit is about craters of the order of 500 meters. The distribution is different in the smaller sizes; there is considerable confusion as to exactly what different processes are operating. It appears that the processes that produce a steady state distribution, from a production distribution that Dr. Shoemaker has drawn there, may not be operative on Mars to the same extent.

The larger craters which Dr. Margulis was referring to, likened to the primordial structure of the moon, are much larger: They are 20 kilometers or greater in diameter. These are completely different scales and a complete dichotomy both in their morphologies and in the structures of the distributions themselves. It is difficult then to compare these directly with the small Martian craters. Because of the nature by which the Mariner 6 and 7 photographs were taken, it is difficult to understand exactly what the distributions are in slopes and morphologies.

RICH: You expect them to be different because of these large smooth areas on Mars which I think have no counterpart on the moon. At this resolution may it be that the maria on the moon would appear like the smooth areas seen in Mars but not be so well circumscribed?

WASSERBURG: I think the greater smoothness must come from the boundary areas.

HOROWITZ: On Mars there is some smoothing process, erosion or something, that acts more vigorously than on the moon.

SHOEMAKER: If you put an upper limit on the abundance of craters per unit area on Mars, it is well below that which you would see for craters that you ought to be able to resolve. The analysis of the crater distribution has been done elegantly by Soderblom. He has developed a closed-analytical model for the interaction of craters inferred to have been formed. Extrapolating distribution above the steady state limit down to smaller craters along this dashed line, you assume that this is the distribution of craters that have actually been formed on the surface—then you predict what the steady state distribution ought to be. It is not a very simple problem; Soderblom has solved it.

ORGEL: What process do people who know about it take for granted that establishes a steady state from the initial distribution?

SODERBLOM: There are quite a number of processes. For instance, Marcus in 1964 posed the idea of crater overlap and, more recently, blanketing by ejecta of distant craters.

The distribution in craters in steady state does extend from a very sharp, fresh looking crater with very steep slopes progressively down to smaller mold-shaped craters. The density tends to increase as you get to shallow craters. We need a continuous model of erosion which tends gradually to decrease. The saturation of surface craters is only a few percent and in the Marcus model you may have a much higher saturation.

The process that we think is operative is one by which smaller craters transport material downslope on larger craters as the ejection processes throw material in conical sheets. The form of the ejected profile is distorted by gravity in this direction, so one can compute what the displacement of the center mass is for each of these small impacts.

RICH: What do you assume for an angle of incidence?

SODERBLOM: Experimental evidence indicates that the axis of this cone of ejection is normal to the surface irrespective of the slope of the surface and irrespective of the angle of incidence except for very, very grazing impacts which are a degree or half a degree after the surface and then funny things start to happen.

HOROWITZ: It is a gas explosion that causes the crater, is that right?

SHOEMAKER: No.

SODERBLOM: There is a shock that starts out in the material which is spherical in form and rarifactions that build up behind that front that are responsible for peeling the material back out of the crater.

SHOEMAKER: The initial ejection of very high-velocity material is very strongly influenced by the angle of impact, but that effect applies only to a small percentage of the ejected material. As the shock front is propagated out into the target medium, it becomes more and more nearly hemispheric and the flow of the material and the rarifaction behind it becomes essentially symmetrical in form to the surface. Thus 95 percent of the material goes out roughly symmetrically around the surface and has only small bias due to the impact angle.

SODERBLOM: The high speed jump as the projectile strikes the surface and the rate of these materials is many times greater than the impact velocities. As the shock wave begins to expand into the material, it expands nearly spherically—and since the rate of shock is not the same in all directions for any isotopic medium—the flexure of this shock front doesn't depend on the angle of incidence.

The axes of the conical sheets are about normal to surface irrespective of the incidence. It is possible to compute the form of the ejected profile by summing these up for all projectiles and by drawing an ejection curve, it is possible to estimate the changes in the topography of a crater with time.

In this model only impacts whose effects can be averaged effectively over the surface are included. A very large impact that takes a chunk out of a crater wall is not included and you can't average its effect over the whole crater surface, but so long as this maximum size which is included in this continuous erosion process remains small, compared to the size of the crater it is eroding, the model will work.

The ratio of maximum size eroding to size eroded is a function of the slope production curve and, for instance, in the case of the lunar uplands, it appears that the slope is much lower than it is by comparison with the slope in the production curve we are talking about.

RICH: After your calculation you end up with a time dependent profile, a weathering with time. Isn't it a sensitive function of the actual distribution of incident energies of the impact, the meteoritic impact material?

SODERBLOM: Instead of scaling energy, we are using a distribution of projections. We are concerned only with the form of the production curve that produces craters at a given diameter. A mass of one may be small but high energy produces the same relative-size craters.

ORGEL: Assuming it is worked out, could you give us some idea of the time for the disappearance of the craters? Is it observable and how does it depend on size?

SODERBLOM: Rather than time, I would say the integrated flux that falls on the surface is about linear with the crater diameter and depends on something between 1, the diameter of the first power to 1.2 to 1.3, depending on exactly the slope of the production curve. I think Ross (1968) came up with a similar result from a numerical and theorized model of destruction on the surface and got about the same dependency, age or accumulated flux with time.

WASSERBURG: If you have strong crater overlap, the immense difficulty of treating this problem theoretically is so great that it is hard to believe that you can really do this.

SHOEMAKER: Do you mean repeated cratering?

WASSERBURG: If you have very high rates of cratering, including even a steady-state case, and are obliged to talk about overlaps of high density, I don't see how the problem is *do-able.*

SHOEMAKER: But this is exactly the problem we are talking about and it is solved.

WASSERBURG: I find that very hard to accept.

SODERBLOM: Pristine, fresh craters overlapping one another is an extremely difficult statistical problem. We are not solving this problem in a discrete sense of statistics of individual craters but treating a quantum effect of the smaller craters which can be averaged meaningfully over the surface as a continuous effect.

If the craters we average turn out to be half the diameter of the area they erode, it is not meaningful. That is one of the stringent requirements.

SHOEMAKER: Let me clarify what the model leads to. The transport of material is such that most material in a crater of a meter or larger stays within the crater, only a small fraction goes out. Mainly the center of the mass of the ejected material is displaced and there is downhill transport. This is a function of the slope which is changing with time.

Larry has solved this problem. Material has moved, progressive profiles start to look like this (Fig. 1).

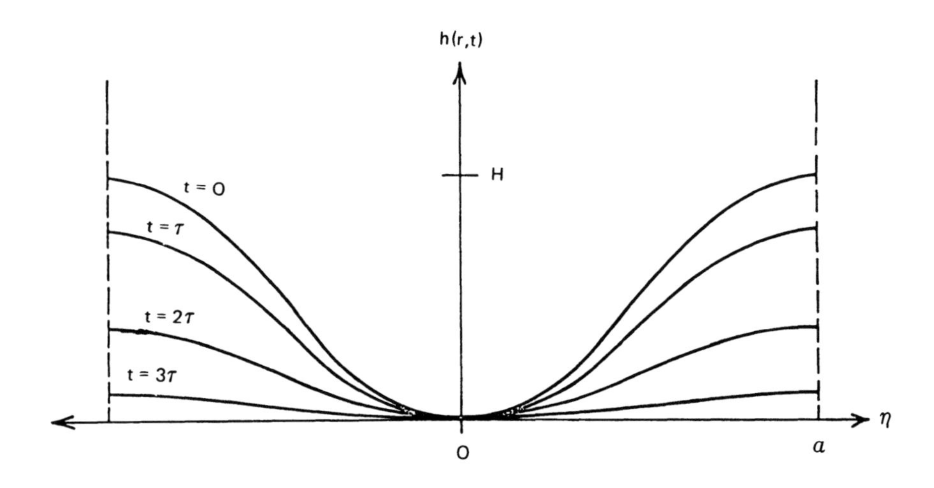

Figure 1. Profiles for increments of time.

They become gentler and gentler until finally that crater becomes unrecognizable. Most material is locally derived but some is thrown away from the region and some comes from a more distant area. From this production function we produce the slope of the steady state curve exactly and we predict its constant very closely. We get a very close numerical fit to the number of craters in a steady state distribution and we can test the model in many other ways.

Another aspect of this model is that you then look at the formation of craters which dig down, and the connected floors of the craters that have dug down form the base of the debris layer. The thickest point on this debris layer will simply be the bottom of the largest crater that has just disappeared in a unit area. The original depth of the crater whose diameter is at the upper limit of the steady state distribution minus its rim height, and all the rest of the craters–if you take the next floors–are the ones that have dug down to bedrock and then are smaller than that and by integrating areas, craters formed under the production curve down to the point as you go to smaller and smaller, you get finally to little craters that no longer can excavate through this debris layer. A model can then be constructed for the frequency of distribution of the thickness of the debris layer. This can be compared against the various means we have of empirically observing thickness and the fit is very close.

Comparison with the observed distribution of craters at the Apollo 11 site predicts a median thickness of the debris of about 4 meters. Incidentally, the arithmetic-mean thickness from this predicted frequency distribution is about 6.5 meters.

WASSERBURG: Suppose you want to deduce thickness from a hell of a big crater, from the biggest crater in the area, the deepest crater which you see, is there a variety of numbers which would come out?

SHOEMAKER: And it has grown out of a big pile of debris.

WASSERBURG: You would have had penetration and mixing, say to 2 kilometers depth, and from this calculation you would end up with still a different meter thickness.

SHOEMAKER: If you look at a 10-kilometer crater, it turns out that about 95 percent of the ejecta falls within the diameter from the crater rim.

Where is the volume in either natural or experimental craters? Part of it is in the ejected rim material, and part of it is formed by flow, mostly plastic, behind the shock, so that the rock walls of the crater are actually moved and lifted up, and the actual rim height is due to the uplift of the bedrock–for example, craters on the earth and experimental craters, the size range is of 100 to a few hundred meters. Only a part of the rim height at the higher parts is due to the layer of debris. However, a significant fraction of the volume of the crater is in a flow spread out in a layer, about 95 percent is a distance from the crest of the rim equal to the diameter of the crater.

In all the time since the mare surface has been formed, it has not been filled again and all craters of kilometer diameter are still craters.

ORGEL: Would you say more about that? That puzzles me. These are in steady state?

SHOEMAKER: That is obtained by fitting the steady state curve which is the same everywhere to the observed production curve at that point. At the Apollo 11 site, craters larger than 141 meters–on the interpretation I am giving–are all there. None have disappeared. Some close to that size have been largely filled in. But no craters of a kilometer diameter have been largely filled in, they still have most of their original depth.

ORGEL: From the size of the biggest ones not filled in, or rather the smallest one you can see, can't you get something similar to time: an integrated impact?

SHOEMAKER: Yes, the surface can be dated from these larger craters. The ejecta thrown out forms a local rim deposit, stuff that has come from greater depth but still a recognizable rim deposit.

WASSERBURG: But when you come out with 5 meters–that 5 meters is not for this crater?

SHOEMAKER: That was, yes.

SODERBLOM: I see the confusion here. When these are measured, some are conceivably 100 meters deep. How do you unravel that from the average of 5 meters?

WASSERBURG: Yes, right.

SODERBLOM: Let's consider that size of crater which is as deep as the average regolith. The total surface area covered by craters between this size [pointing] and about twice that size, is about 70 percent. The number of craters decreases very rapidly with increasing size. There are many of these much deeper and irregular craters. They are very, very small and they cover a very, very small fraction of the total surface by comparison to this average size which is blanketed in surfaces many times.

SHOEMAKER: Let's try again. Each crater on the moon goes through this sequence (see p. 38). It starts out in some original shape and then evolves in shape until finally its form is so low we no longer recognize it as a crater in the pictures. There is a crater within the reference area that we have counted that has gone all the way, so that as far as its photometric recognizability in the pictures goes, it is flat. In one of these unit areas of 100 million square kilometers today that crater is filled in with debris and is gone. The thickness is the height from the surface, the original bottom of that crater, and this height is back to the original average surface level.

Four meters is median thickness of a calculated distribution and this thickness is about 22 meters. This is the expected thickness of a level surface where a crater has been formed and subsequently it completely eroded away.

If we go to the larger craters, their evolution in form gets less and less and you can scarcely recognize any changes taking place at all in a 10 kilometer diameter crater in proportion to the shape of the crater.

The oldest craters will also have a thickness of debris in the bottom, not exactly 22 meters but a little thicker because the transport tends to fill up the bottom very quickly. The oldest or most subdued craters whose thickness will exceed 22 meters, maybe 30 meters or 40 meters, are mainly made up of material derived from the crater wall. There is a subdued rim around it and material is transported out away from that rim.

Most of the material that is involved in its formation is still in the vicinity of the crater, partly thrown out and partly moved back in by repetitive, small cratering events. These local things are not considered in the frequency distribution I talked about, which is for a surface that has been filled up level again. We ask, if we look at the lunar surface and stick probes down through it now–actually the probes are little craters that have been formed very recently–what is the expected frequency distribution of thickness of debris at these sites? I am trying to solve this. It ranges from a maximum of about 22 meters down to a minimum of less than a meter–a somewhat artifically defined minimum that I use to make the calculation simpler. That is the range; the distribution is very strongly skewed. In most places the thickness is less than the average arithmetic mean of 4 meters. The average is 6.5 and the maximum is 22 for that distribution.

KAPLAN: What would you expect it to be in the highland area?

SHOEMAKER: The calculation is for a level surface. All bets are off for a surface with big relief and large net transport, but for a level surface, a mare, the calculation worked well for about 5° tilt. For example, it works well on the slope of the crater Tycho. On this model the thickness is expected to be probably three times as deep as the predicted thickness, median thickness of the regolith at Apollo 13.

MURRAY: What is the issue here? I am not involved in the lunar work, but I am interested because of its application to Mars. I have not found any reason to challenge the basic approach. What contradiction in the data makes you wonder why this model of the regolith thickness isn't a reasonably accurate description?

HOROWITZ: I think the issue is just to understand the model.

MURRAY: But it is a very detailed thing and many people have been working on it for a long time. What you can observe, you observe. There are other ways of getting regolith thickness besides calculating it or besides measurements made *in situ* by astronauts. It works reasonably well. My criticism is that it isn't worth all the trouble. It explains adequately that it is a mixing layer, giving you a feel for the age and how it varies in thickness, and so be it. Although in an absolute sense, in any one place on the moon, it may depart significantly, in a statistical sense it is a good model.

ORÓ: I think it has been a wonderful learning experience.

SHOEMAKER: I am talking about the contact of the loose debris on bedrock. Everywhere on the surface the material has been turned over once to this depth, and it has been turned over more frequently than once by smaller craters as you go to shallower and shallower depths.

KAPLAN: Gene, from your model is it possible to calculate the maximum distance from which any particle may originate at any one point?

SHOEMAKER: I can give you the statistical expectation. The maximum distance a particle can go turns out to be the circumference of the moon.

One can extend this argument with very small craters, based on a very long extrapolation from craters larger than the steady state limit, tiny craters, cutting the distribution off at 10 microns. There are independent reasons for thinking this extrapolation is approximately correct. From this, then, you can compute the volume of material ground off a rock by very small craters. The probability that any single crater will be formed on a rock that is large enough can be computed. The effect is just to smash the rock into small pieces.

It turns out that the mean-lag time of rock fragments over the complete size range is somewhat greater than the time of turnover to a depth equal to the size of the rock, equal to the characteristic dimension of the rock. If these calculations of the model are correct, it means that before a rock is ground away or destroyed it will become buried at least once. In fact, the time to turn over to a depth equal to the dimension of the rock is about half the lifetime of the rock in various size ranges. A large majority of rocks therefore are expected to be buried at least once before they are lost. If the rocks happen to get thrown into an already existing crater, some have a very high probability of getting buried deeply enough so that they stay a long time before they get exhumed. If you use this to compare against exposure ages, then some exposure ages are expected to be much larger than the mean prediction here.

If I use Wasserburg's figure for the age of the lavas, the average expected exposure age for a 10 centimeter rock is about 15 million years, that is, considering it is thrown out on the surface and in most cases will be buried again and then thrown out on the surface again.

WASSERBURG: Down to roughly 5 or 10 meters?

SHOEMAKER: No, no, in most cases it will be buried to a depth only about twice as deep as the size of the rock.

HOROWITZ: What is a comparable figure for rocks that are exposed on the surface of the earth that disappear and get buried?

SHOEMAKER: The problem doesn't apply to the earth. I don't understand. You mean a fragment thrown out on the earth?

HOROWITZ: Twenty years?

SHOEMAKER: No, they last much longer. This is a problem I am interested in working on in the Colorado plateau. This is very dependent upon climate. A sort of horseback guess is that the typical lifetime of a rock fragment in a relatively arid region–for a 10 centimeter rock is of the order of, say, 10,000 years.

KAPLAN: This depends on whether you mean an isolated fragment or a fragment as part of a larger rock.

ORÓ: I don't think it is a compatible situation.

KAPLAN: I think before 10,000 years are over the rock would be weathered or washed away.

SHOEMAKER: I am basing my estimate on the rate of weathering of surface features over a 100-year time span.

MURRAY: There are archeological data that show that desert pavements hold up for thousands of years in the Southwest United States.

SHOEMAKER: The spread in age is enormous.

MURRAY: There are the implements all tied together that primitive man used right there and there are some fossil surfaces that do last for several tens of thousands of years.

SHOEMAKER: I think you could find much older individual rocks.

MURRAY: It is extremely variable.

SHOEMAKER: In this case age is extremely sensitive to the composition of the rock. Cherts and other materials that form archeological artifacts can last an enormous length of time.

HOROWITZ: Because there is a lot of chemistry in the surface of the earth.

ORÓ: I don't think it is a compatible situation.

SHOEMAKER: You wanted an idea of how long a rock can sit out on the earth's surface; the typical case is orders of magnitude.

RICH: The real question we want to ask is how long this process takes place on Mars. Maybe we can discuss that tomorrow.

ORÓ: What is the thickness of that mean average of turnover of regolith? It certainly is important to those of us interested in investigating life there.

MURRAY: We have some indirectly relevant observations.

SHOEMAKER: The actual exposure range–calculated from spilation products in the rock fragments brought back–is based on the helium isotope in the rocks from an age of about 10^4 to a million years. In a broad way this is consistent with other measurements. If I read the papers correctly the average exposure age looks a little higher than my calculated mean.

ORGEL: Again for those of us who don't understand this, can you tell us what "mean age" means?

SHOEMAKER: In the case of the spilation ages this means the residence time of rock fragments–not soil–within a meter of the surface. That depth varies depending upon what you are looking at, generally it is much closer to the surface.

ORGEL: It has to be present in an integrated piece of rock within a meter of the size for that time?

SHOEMAKER: Yes. We can also try to estimate the ablation rate of rock due to small particle bombardment, as Walker (Crozaz et al. 1970) did, mainly on the basis of particle track studies. These tracks probably are made in part by solar flare particles. He estimates an upper limit of the grinding down rate of the order of a millimeter. Extrapolating down the crater production function, I actually get 2 millimeters per million years. I consider it good if I am within a factor of 2 with that extrapolation.

ORGEL: What do you assume about the rate of production of rock? You must produce these rocks from somewhere.

SHOEMAKER: New coarse crystalline rocks are formed each time a crater digs down through the regolith and brings up new bedrock.

RICH: The new rock is formed in the material that is underneath the crater through the impact.

SHOEMAKER: About half the rocks are crystalline chunks seen on the surface and about half are simply remade instant rocks.

RICH: And these are in the part that is in the bed of the crater?

SHOEMAKER: Yes, and each new crater that penetrates through the regolith will have a rim of coarse rock fragments. West Crater, the one avoided by the astronauts during the landing, is a beautiful example. About 30 meters deep, it has an ejecta rim of very coarse debris and rays of coarse rocks going out from it, from where most of the new crystalline rock fragments on the surface have been buried once and re-exhumed in the fine grained debris; the calculation of the lifetime just tells us how long a rock lasts once thrown out on the surface, how long it is likely to have been there. The thickness of the regolith now is great enough in most places to effectively shield the material from cosmic rays, at least to shield it from the effects that would go into the exposure-age calculation.

ORGEL: I presume the actual abundance of rocks on the surface is changing with time.

SHOEMAKER: Yes, in an interesting way. Very young surfaces have a much higher abundance of rocks. The reason is that the spacing of craters that excavate fresh rock is much closer–so it is due actually to the presence there of the regolith.

ORGEL: Can you use the abundance of rocks to estimate the depth?

SHOEMAKER: Yes, in a gross statistical way. You can indirectly get the thickness that way. The calculated thickness fits very well the observed thickness of the regolith and the abundance of coarse fragments.

I want to discuss the structure of the fine grain material. We had a good look at this with Surveyor. This trench into the surface was dug by the footpad of the spacecraft as it slid down the wall of the little crater. The wall collapsed leaving the lumpy structure of this material. The wall here [pointing] shows how the fine grain material tends to be compacted into very weak, soft aggregates that are easily smashed if hit with a scoop of the Surveyor or if walked on. The footpad of the spacecraft spanks them down and smooths them out, they are very soft lumps. Here is one lump apparently made up of still smaller lumps, so there is a very weak aggregate of clod-like structure to most of this material.

Here we see the surface on the rim or upper wall of the crater in innercrater areas. In some microbreccias brought back, there is a thin but perceptible bedding. The material to be bedded is expected where regolith material is accumulating in craters. The rate of buildup of material there is greater than the turnover rate, so that individual sprays of ejecta from small craters lead to a bedded structure which we actually see. It is very crude, not a well formed bedding like we see in water laid deposits on the earth. Nevertheless, it is a distinguishable layer in some returned specimens of regolith.

WASSERBURG: How long should a bedding layer last?

SHOEMAKER: I don't know quite how to answer, it is a function of the scale in which it is broken up. I am hypothesizing that the bedding layer at the scale of a rock fragment brought back in a microbreccia is formed by the filling up of a crater substantially greater in size than the fragment we see. And the bedding layer, or a fragment of it, can last as long as fragments of the microbreccia.

KAPLAN: How much sorting is there in that layer?

SHOEMAKER: Observable, but very weak. There is a distinct planar structure in some microbreccias and the entire basis for believing that it is bedding is that there is a select sort of coarse and finer grains from one layer to the next.

KAPLAN: Repeated cycling type?

SHOEMAKER: No, not really. If you see a few cases, you will note that their layers are slightly coarser or finer grain.

KAPLAN: It is not a cyclotherm?

SHOEMAKER: No.

SCHOPF: Isn't the bedding defined on that basis, not in terms of segregation of rock types?

SHOEMAKER: No, no.

SCHOPF: Or glasses versus fragments?

SHOEMAKER: No. Actually, lamination with a characteristic thickness of about a millimeter is seen in some of the microbreccias. There may be coarser bedding structures. Locally all different coarse layers might be formed by an accident of overlapping ejecta rim deposits, that could have come from materials of different sources.

SCHOPF: How sure are you that these are really bedding plans as opposed to a sort of artifact?

SHOEMAKER: This planar structure in many of the microbreccias is easily seen. We get planar structure just due to shear flow in the rocks–incidentally, in instant rocks most of the time. But the reason for thinking it is not due to shear in lunar rocks is that there seems to be sorting.

SCHOPF: Do you see this in thin sections?

SHOEMAKER: We have looked at the surfaces of the rocks.

SCHOPF: I have looked for this in about a dozen microbreccia thin sections and I haven't seen it.

SHOEMAKER: I have not looked at thin sections. It should be observable in thin sections, but the structure is so weak that you need a large section to see it; in such a little section your chance of finding it is small.

WASSERBURG: I don't think there is any evidence for a real bedding. If you are going to have a mixing model like this which is applicable to any planetary surface where there is impact mixing, then the bedding basically has to be wiped out.

SHOEMAKER: It is only preserved for a short time in fragments like this.

WASSERBURG: Right, there is a life time for the thing that is bedded.

KAPLAN: Gene, you said very thin beds, millimeters thick. If there is continuous mixing, whether millimeters or centimeters, they should have been destroyed.

SHOEMAKER: Yes, bedding will be seen only where deposition rates are greater than the turnover rate, really only in two places on the surface. Mainly, this is seen in the bottoms of craters as they start to fill rapidly just after they are formed, but somewhat in little embankments of material on the sides of rocks that are built up—and those are small areas. Here [pointing] I think we are seeing repetitive stirring and formation of weak clods by shock and then breaking them up and again forming them and breaking them up, bedded material is seen.

One can go back in the laboratory and fire a projectile into a rock target. With large resources and the right framing cameras—with a million frames per second frame rate—the trajectories of particles thrown out can be timed. The amount of mass that goes out at varying velocities can be measured. Doing the calculations you can ask: Where would this stuff have gone if you made that little crater on the moon? This curve is the one hop curve based upon an experiment of firing a projectile at 6.4 kilometers per second into a machined smooth surface of basalt—a hard experiment—and then taking the observed velocity as a function of mass ejected from the crater and then calculating the spread of this material from the crater on the moon. The cumulative mass, about 50 percent of the material, falls within 1.8 kilometers.

ORGEL: Is this averaged over all sizes of impact?

SHOEMAKER: It is one experiment.

ORGEL: I am not with you. Don't you have to know what the impact is?

SHOEMAKER: No. It is almost independent of the size of the crater. This little tail [pointing to arrow] does depend on the impact velocity, but the bulk of this curve is relatively insensitive to the impact velocity of the projectile. This is all the mass thrown out independent of particle size. There is some sorting by size, the coarsest particles are going at lower velocities. I am using these data because they are all I have.

Actually because of the scaling relationships you get the same kind of curves for big craters and also for little craters, not strongly dependent on crater size until you get to extremely minute craters. For craters from a few centimeters up to meters, the curve should be about the same.

This shows the distribution of material thrown out for a single event, calculated for the moon. How many times on the average has the material of the regolith been thrown out of craters? This comes out using the model of turnover that I have been building on since Larry first described it. The total integrated volume of material displaced in craters, equivalent to the crater formed in bedrock, can be calculated. For material moved in velocities comparable to this experiment, that volume turns out to be about 20 cubic meters per square meter on the surface. The average arithmetic-mean volume of the regolith is calculated at 6.7 meters per square meter, so the number of times that the bulk of the regolith has been moved in cratering is about three, in high speed events. Actually it has been moved an order of magnitude more times as very low velocity ejecta, which goes into the calculation of why craters disappear, that that low velocity stuff doesn't as strongly affect the dispersal of material on the moon beyond the kilometer range. The actual dispersal of material is a curve for three hops. It doesn't follow quite the shape of these curves. Actually about 50 percent of the material in the regolith on this model comes from a distance of about 3 kilometers from where it is now. About 5 percent has come from distances greater than 100 kilometers, an accumulation of a little over half a percent–using this impact velocity and this number isn't very good–have actually been lost from the moon reaching escape velocity. This number is highly dependent upon the assumed impact velocities.

There is a bulking problem in the calculation of regolith that I have not included. You start with dense rock and grind it up, it expands in volume, and that is not entered into the calculation. Much material is not lost.

The contamination of the regolith with meteoritic material, the projectile material, looks as though it is small. It is in the order of a couple percent.

RICH: Do you think any of the samples from Apollo 12 were bedrock?

SHOEMAKER: Do you mean were they locally derived?

RICH: Yes.

SHOEMAKER: If the astronauts had sampled the right rocks, we could be almost dead sure of certain rocks. On a statistical basis I think we can say very confidently that at least half are from local craters where the samples were collected, because most of the samples are collected from rims.

RICH: Virtually none is undisturbed bedrock?

SHOEMAKER: Yes, none is undistrubed bedrock. Had the astronauts been able to go down into one crater–Bench Crater–they might have returned with a beautiful little nubbin of coarse rocks in the bottom. Conrad tried, he got partway down the slope and decided it wasn't so smart. But, that would have been a good place to test.

RICH: Do the astronauts understand the difference between the regolith and the bedrock?

SHOEMAKER: Let us say there was an attempt, not trivial, an extensive attempt to try to educate these guys. The Apollo 12 crew did not have nearly as extensive exposure as we had hoped. I think the Apollo 11 crew understood these problems quite well. In retrospect, our attempts to educate were much less successful than we had hoped.

RICH: How about the Apollo 13 crew?

SHOEMAKER: They have gotten a much better background on this than 12 did.

SCHOPF: Why does no meteoritic material have compositions similar to lunar rock?

SHOEMAKER: Let me rephrase the question: Why don't we find samples of the moon among the meteorites? Almost all of the material that is thrown off at escape velocity is melted. T. Gold at Cornell University thinks that all of it is; I think some rocks may be very strongly shocked that aren't melted. But almost all will be melted and we haven't recognized them. They aren't recognized as meteorites.

Almost certainly there is lunar glass. It probably came out as very small particles in the atmosphere, and it may be among the varieties of small glass spheres found in Antarctic ice and various other environments. I don't know how these would be separated from terrestrial contaminants.

RICH: Look for high titanium content.

SHOEMAKER: We probably could find it, now that we know what to look for.

SHELESNYAK: We plan to have lunch in about 5 minutes and I think we ought to have a 5 minute stretch first.

The conference recessed at 12:25 p.m.

Saturday Afternoon Session
February 28, 1970

The conference was called to order by Dr. Eugene M. Shoemaker.

SHOEMAKER: I think our next step is discussion of the hard data rather than the philosophy, so we will call on Gerry Wasserburg.

WASSERBURG: I want to present some data on the lunar samples in a cursory fashion. Instead of a formal speech, I hope we can more specifically discuss biological or organic problems of interest to the people here in terms of data from a more primitive type planet than Earth. I want to first quickly review some of what Gene Shoemaker spoke about to give you a feeling for some results in the perspective of the confusion which exists in real life without, I hope, obfuscating the fact that there are some rather hard conclusions.

The first slide shows a mare and highland areas on the moon. Anybody can tell the mare material and the highland material apart.

The next slide shows lunar areas which are mixtures where the distinction between highland and mare is not so clear. The next slide shows a transition region between mares: some great big clusters here [pointing], a crater here and a mare region, and only minor obfuscation over here. We need this before we get down to the nitty-gritty where we will draw our obvious cosmic conclusions. The next slide shows the nitty-gritty: the real surface and the confusion. This is presumably bedrock, so this clear distinction between a mare site—from a distance this is clearly distinct from the non-mare areas—breaks down in real life: The place you saw this morning is a rubble heap of rocks, all sorts of junk dumped over. The physical origin of these materials is not known. The perspective I want to bring is that from a gross morphologic point of view we see a distinctive unit, yet at the detailed level we are looking at a rubble heap of rocks where the relationship of the individual fragments to the broad morphologic features is almost totally obscured.

SHOEMAKER: Gerry, may I comment? There are well-known variations of color and albedo on the surface, some quite closely correlated, and the boundaries between materials of different color and albedo on the surface in some cases are relatively sharp. It would be difficult to understand those boundaries and those correlations if there were not a close relationship between the bulk of the pile of rock debris and this morphological unit of sort of interlocking bedrock.

WASSERBURG: Yes, that is valid. But you are backing off to a gross point of view from a broad distance. I want to emphasize that once down to the object, the object as a distinct entity loses some of its identity. I don't think this totally obscures the experiment but it does make it complicated.

I will not say we have a piece of rock causing the mare itself and not from a broad rubble heap reflecting some different materials.

SHOEMAKER: We haven't gotten samples of that from Apollo 11. Had there been more time, they could be gotten.

WASSERBURG: That is very possible. So far, without a study involving crater bottoms and a more rational sampling program, the analyses we are going to talk about from 11 and 12–which don't, in fact, exist, they haven't been carried out–basically involve sampling over a terra-incognito which morphologically has an extremely distinct characteristic. There is no question about that. When you are down on the ground it represents something which shows all of the complications which you demonstrated in the problems of the formation and mixing of the regolith.

SHOEMAKER: Nevertheless, there are systematic surface patterns which vary in both albedo and color, and have sharp boundaries between them at the telescopic scale, although they are diffuse at the scale of a few kilometers. There probably is some correlation between the material mixed up in that debris and what is locally underneath. When I say local, it would mean within an error of some precision of the order of 2 kilometers.

WASSERBURG: I agree. The point of emphasis here was that things are grubby and confused. I have one little rock that happens not to be in the picture. It could be any of them. The game gets a little hairy because we now have an isolated individual rock that is unrelated to anything except that they landed on that spot. The sample they got is this mixture of fine and coarse particles from a variety of locations and sources, which are not really known. I will try to point out later that besides the instant rocks, in terms of the larger trace, we may be looking at a total of only two rocks from Apollo 11.

LUNAR GLASS

WASSERBURG: This cratering [pointing], of course, takes place on a much finer scale, too. Here are several craters and here is part of the glass sausage which has been shattered by a high velocity impact and busted up. This shows small glassy particles of crystalline materials in the regolith, which also appear in the instant rocks.

RICH: What is the diameter?

WASSERBURG: About a millimeter. We will get down smaller.

The particles in the soil are glossy crystalline fragments reflecting the impact characteristic that Shoemaker talked about, down to all scales. Here is a droplet of glass which perhaps formed and fell back in when it separated and perhaps, with strain and instability, it was blasted back on itself. That is a half millimeter, a monstrous object. There are glasses–in answer to the question that was asked before–which were truly molten. They were similar to when you blow glass and it got quenched so that they dropped below the point where their viscosity was high and kept in fluid form. Some show obvious surface tension features, instability.

KAPLAN: From your earlier statements. Gene, I had the impression that you felt that most of these glass spherules were derived from meteoritic impact on the lunar surface. Is this correct?

SHOEMAKER: Yes. Do you believe this also, Gerry?

WASSERBURG: I will tell you what I believe in a moment.

These [showing a slide] are some of the glassy objects. This one has a hyper-glossy impact area, and is full of mountains down to finer scales of 50 microns. It is actually half a glass ball and you are looking at a ring across it. It has been fractured. This is small, about 2 mm. in diameter and was broken in half. The broken half was cratered by hypervelocity impact of small particles many, many times, with a surface crater density of around three craters per square mm. This is such a crater here. Some objects must have hit it first to break it. Another object hit it and splattered glass all over the glass and other objects hit it and cratered it again on a smaller scale.

RICH: Is there any opportunity to trap the particle striking the surface?

WASSERBURG: It is mostly gone. Shoemaker predicted it and he is right.

RICH: Is it volatilized during the impact?

WASSERBURG: I will try to be specific. We have been unable in any of these craters to find the material which made the crater.

HOROWITZ: By "hypervelocity" do you mean escape velocity?

WASSERBURG: Ten to thirty kilometers per second.

SHOEMAKER: You can make some statement based on the melting. All you can say about impacts is that they are at least several kilometers per second but if you look at the actual frequency distribution of encounter velocities of photographic meteors, the median encounter velocity is about 25-30 kilometers per second. Probably we see velocities in many cases at least this high, tiny particles that make the crater.

HOROWITZ: Following up Alex' question, are there low velocity particles hitting the moon–I suppose the minimum velocity is escape velocity–that may be preserved? Suppose there were a shower or constant rain?

SHOEMAKER: More of that has a chance of being trapped on the surface than high-speed stuff.

WASSERBURG: Of course. I believe you actually see things that are trapped material but at other places there is no evidence.

Here is a closeup of one crater on a feldspar crystal that has isostasized–it is glass and molten inside. It is surrounded by brittle fracturing all the way around, so, you see the inside of one crater as I just showed you at a greater distance where the central part of the crater is made totally molten.

All this [pointing] is moon dust just stuck here. If it's cleaned out, the surface will be infinitely smooth, with black marks and bubbles on it. Inside there are broken fragments of other glasses plus more glass spherules or bubbles. That object is a glass bubble. An Englishman carefully described it as something like flue or fly ash that comes from factories. It is bubbly and gaseous as heck. All are full of gases, most glasses are full of gases. This is half a glass ball, pretty smooth. This is a busted bubble.

RICH: That is a hemisphere.

WASSERBURG: It is actually a full sphere, which in my infinitely meticulous handling–I don't do microbial surgery but I am pretty damned good –I busted in half. It was a total sphere.

As Gene has pointed out to us, when these craters are made they are essentially jetting out material–as if you took a garden hose and put it in something and it was just squirting or blowing the junk back out.

This is the back trail. We are seeing lobes and drops from instability and a trail of this stuff which is just dishing out in rays from the center down to a very small scale.

YOUNG: What size particle will do that?

WASSERBURG: Probably 1 micron.

SHOEMAKER: Typically, you can figure that the projectile will have been probably one-tenth the diameter of that inner glass lining.

WASSERBURG: It depends on the energy considerations.

RICH: If you make that assumption, can you make some guess about the speed?

WASSERBURG: It depends upon your efficiency.

RICH: Density?

SHOEMAKER: That is not a useful way to get information because that is how I made the assumption about the speed.

WASSERBURG: One-half mv squared equals the amount of energy to get the stuff out and the question is, how much energy must be made to get rid of the material? This can be done by flow, evaporation, or whatnot, and in any case you get melting.

I didn't mean to concentrate on this but I didn't know who the speakers would be. I guess I didn't read my mail.

RICH: You want to end up with something having vague biological relevance.

WASSERBURG: That's right. The point is, when we talk about the soil there are all these kooky little particles; this aggregate makes up the soil. These were all molten silicates above 1000°C.

KAPLAN: I would think that particles that came in with such velocity, on striking a hollow sphere, that the hollow would have shattered it.

WASSERBURG: That is not true. That experiment was carried out carefully since we have a hollow sphere which is covered on the surface, but it rarely is punctured even with 50-micron craters. A 50-micron diameter crater does not puncture a 50-micron wall.

RICH: Was this hollow?

WASSERBURG: That was solid but there were bubbles on it, as I showed you before. In fact, this is a little teeny bubble. There are lots of little bubbles in here. So, this is a glass which was impacted to make another glass and this is a brittle fracturing around it.

KAPLAN: Gerry, do you mean it does not in fact puncture it or that there is a healing effect?

WASSERBURG: No. This is pure brittle fracture so we did that experiment. For example, here is a crater that hit and here is a cross section of it, so you have a shatter zone going down inside like this, and then you have this melted *gilch* coming out. It tapers off, it didn't go through.

SHELESNYAK: Do you have any indication of internal pressure in the hollow spheres?

WASSERBURG: Yes, there must have been pressure to support the bubbles but the pressure could have been negligible depending upon the dynamics, the viscosity of the materials, and the temperatures at which they were molten.

In addition, on this glass, if you look carefully you find these lovely sexy craters of glass and these busted objects, things which I call the eye of God. They don't photograph very well. They are actually glass bubbles in various degrees of spawling off, depending on the thinness of the wall and, in fact, they get to be totally just bubbles on the surface and the bumps on the surface with this removed, so here a 10 kilovolt electron beam can actually penetrate. It is a typical kind of glass, just frothing full of bubbles.

HISTORY AND AGE OF THE MOON

WASSERBURG: There are many phases which can support such a bubble, namely, the decomposition of the silicates, as Arrhenius pointed out, is more than adequate, more exotic compounds need not be looked for. All of these objects are saturated in the solar wind, so if this stuff melted then bubbles, composed mostly of helium, would also be made.

HOROWITZ: Why mostly helium? Why not hydrogen?

WASSERBURG: It would be both hydrogen and helium; helium is at the very highest concentration and is a competitor with hydrogen.

HOROWITZ: What happens to the hydrogen?

WASSERBURG: Hydrogen doesn't stick quite as well. These gases are now in a funny little layer about 500 angstroms thick. I don't think these phenomena are critical for Mars. These come from the solar winds, with a typical particle flux J of around 10^8 protons/cm^2/sec. The total surface has been exposed to several mols of hydrogen/cm^2. The place to find organic stuff is here because the density of protons is so thick to almost be solid hydrogen except it is in steady state and it also has carbon. In this company we may talk about a hydrogen to carbon ratio slightly greater than 10^3. This is probably the place to look for natural hydrocarbon compounds.

SHOEMAKER: Is that mols percent or ratio?

WASSERBURG: That is numbers.

ORÓ: But then you have to multiply these ratios by the sticking factor.

WASSERBURG: Yes, I don't know the sticking factor. The carbon will stick a lot longer than the protons.

SHOEMAKER: It might be down around 10^2.

WASSERBURG: Probably the most important place on the lunar surface to find complexes pertinent to primitive organics or hydrocarbons–and not just due to the water-gas reaction, which is mostly what people are finding in these sophisticated experiments which even Nobel laureates carry out–is in the outer skin and layer of every little grain, each of which has sap. Each little scoop of particle has sap for circa 1,000 years at the free surface of the moon.

SHOEMAKER: Isn't that a function of particle size?

WASSERBURG: I mean for the fines.

SHOEMAKER: That is, for particles with a mean size smaller than a millimeter?

WASSERBURG: More or less.

KAPLAN: Gerry, can you make any theoretical prediction of what the H over C ratio would be?

WASSERBURG: Theoretically, it depends on the sticking factor and sputtering characteristics of the surface, et cetera. I would expect an H/C ratio significantly less than a thousand.

KAPLAN: Isn't it in fact about 1:1?

WASSERBURG: We are not talking about the bulk sample, but about 500Å skin on every dust particle. If you are going to take one pound of lunar dust to do the experiment, you are looking at the volume and not surface. My point here is that the surface of these grains is saturated with particle bombardment of hydrogen, carbon, nitrogen–in roughly the appropriate abundance. Giving it to Stanley Miller, you might get an amino acid out of this thing.

MILLER: Glycine.

WASSERBURG: Hydrogen has been identified by several workers. The most clearcut H/C ratio came from Epstein and Taylor. Their ratio shows that if life is not too complicated, there is mostly deuterium-free hydrogen which ought to be in the sun; otherwise, it is going to be pretty embarrassing to burn stars and do some deuterium over.

SHOEMAKER: Extract hydrogen after getting rid of probable contamination.

WASSERBURG: Yes. The H/D ratio was measured from samples in which we tried to separate out contamination. The solar wind stuff that is stuck in the particles divided by H/D on the earth is greater by 10^{10}. This is solar wind material from which deuterium is stripped.

KAPLAN: Wasn't 10^9 to $10^{8.8}$ the value?

SHOEMAKER: It depends on where you sit.

WASSERBURG: It is 90 percent depleted in deuterium. They actually give it parts per million but I have never understood that. Deuterium is depleted in materials brought back from the moon by a decade compared to what we find in the earth, so it is all consistent with what we know.

When I say *surface grains* it is very important: It means every grain on the surface of the moon has been totally exposed to solar x-rays and to particle bombardment, not shielded at all. From Shoemaker's predictions and our work it is clearly about 5 meters deep.

[Drawing] On a coarser scale, imagine a grain or a rock. What does cosmic radiation mean? The problem needs to be reformulated, but anyway, first we have meteorite cosmic rays: the so-called undressing problem. With a great big object like this, the galactic cosmic rays come in and penetrate to a depth of about 2 meters. If the object is broken up, and if it has a big volume and small surface so that it is mostly buried material, then the material that is buried to a depth of 2 meters never would see cosmic rays. The beginning of exposure is only when it is yanked out or broken up.

We may now figure out how long a large object has been exposed to cosmic rays. With small-sized target production rates the exposure rates of most meteorites have been determined. Arnold and his co-workers at La Jolla and Perkins and Bencke and a large number of others have come out with similar results. The typical time that a rock picked up on the surface of the moon from the rock pile has been exposed to cosmic rays, ranges from around 200 to 500 x 10^6 years. This won me one dinner and a bottle of champagne.

The surface objects we see now have, of course, been buried down deep or at shallow depths and have then been brought back up. They could not have been sitting there for half a billion years. Since these objects were found they had to be exposed to cosmic rays, but before they were buried down maybe 500 centimeters, maybe 20 centimeters, maybe 1 meter or 20 meters deep. They may have been beaten on for 500 million years and they could have spent half their lifetimes up and down the first few meters in the deep lunar soil getting shielded.

Other objects are on the surface for only a short time. Right on the surface rock they have had different periods of exposure to cosmic rays. These then must have been dumped out more or less recently.

From how deep are we sampling dust? Forget the rock, it has a 500 million-year exposure age but has been up and down. A simple calculation, not involving crater counting, requiring the rate of production of certain nuclei, can be done. For example: The production rate of xenon 126, by high energy reactions of the protons on rare earths that make bits from m.e.v. particles is known.

The moon, like Mars or any of the planets, does the following thing: The cosmic rays have only limited penetration and we roughly know, as I will show you, the age of these materials is 4×10^9 years. So, we count how many xenon 126 atoms we have per gram. Having counted them, we say if they were in the zone of cosmic ray penetration, the production rate would be zero.

It is not strictly true, because it is an exponential thing, but it is O.K. Knowing the production rate, the only way I can get material on the surface–to show this low concentration of xenon 126–is by my mixing it to a depth of 4 to 6 meters. Thus from just the galactic cosmic ray production we calculate the mixing depth of the lunar surface.

There is a whole cascade of such experiments one can do, including sophisticated ones with neutrons.

The maximum depth of mixing expected from the neutron data is 100 meters. When this calculation is done for a moon without hydrogen and with titanium, it will certainly drop by at least a factor of 3, if not more. Forgetting about the detailed agreement between the types of Shoemaker's calculations and my calculations and observations the basic issue is if the mixing goes to a depth of around 5-7 meters, certainly 100 meters is a super upper max probably too high by at least a factor of 3.

This slide shows all of the high quality potassium and rubiduim ppm data published on lunar samples, my data and that of my colleagues and one or two people we think are good enough. That is the way life goes. These are lunar samples. These are low potassium rock, high potassium rock, instant rock and soils.

Looking at all the rocks brought back, from a chemical and mineralogical point of view—besides the instant rocks—you can see they form two coherent groups. Really two rocks were brought back, this is exhibited in every characteristic. The ppm rubidium versus potassium is one such manifestation, and I could do this for eight elements.

The soil, if totally local, must be a mixture of these two rocks, but I may never mix the rocks and come out with soil. The difference looks slight. They look correlated. They are not. This is true for every element. If you draw a line between the low K rocks and high K rocks on any diagram at all, the soil is off. Instant rocks look like a mixture of a fine-grain material that I call soil and the high K rocks. The implications are that breccia are associated with the high K rocks.

Now I would like to return to the problem of impacting projectiles. Some glasses which are present in the soil and in the instant rocks have FeNi balls. The Fe is 90 percent iron, 10 percent nickel. Most meteorites have metallic iron nickel in them. Therefore, it is probable that these are the residue of impacting particles.

There is Fe metal in all the lunar rocks and no nickel, so this has rather profound implications, suggesting these are projectiles or parts of projectiles distinguishable from indigenous lunar material.

A profound observation concerning the presence of water and reducing conditions can be drawn from the iron metal in the rocks themselves. Although unnoticed in the reports of Arden and Albee and Bryant Skinner, the people who first identified iron in the Apollo 11 samples, iron metal does not occur simply as iron metal but is associated with ex-solution blobs in iron sulfide. The iron wasn't originally there as iron metal but is iron metal separated out of an FeS melt and so it is not primary. It is my observation on Apollo 12 that the iron metal occurs separately. These two statements are significant differences.

KAPLAN: What do you mean, it is not primary?

WASSERBURG: It is ex-solution from an FeS melt.

SHOEMAKER: It is an unmixing while still liquid.

WASSERBURG: It is still liquid.

SHOEMAKER: They are actually little spherical bodies.

KAPLAN: I don't know why you ascribe to it the word *primary*.

WASSERBURG: Primary to the extent that there was not a melt with molten iron. In one case it did not make FeO and in the other case it never had the chance to make FeO.

SHOEMAKER: The free metal phase became a free metal liquid after most of the rock had crystallized, because the sulfide matrix in which these things occur is already interstitial bodies between the silicate crystals.

ORÓ: What is the difference between Apollo 11 and Apollo 12?

WASSERBURG: My statement about 12 is an opinion-based one on a few minutes' observation. In Apollo 11 they were not separate pieces of molten iron; they were pieces of iron dissolved in FeS. In Apollo 12, they may have been molten iron. This is just a preliminary observation, but I wanted to warn everybody.

On the earth you almost never find metallic iron in anything except in cases like the Disko Island basalt where the rock is included in a reducing material like carbon which reduces iron oxide and hydroxides to iron metal. There is no nickel in the rocks which is where the iron would go, presumably.

MILLER: When you say "no nickel," that is less than how much?

WASSERBURG: The nickel content of a lunar basalt is just a few parts per million. This value in a terrestrial rock would be embarrassingly low. That means that the lunar rocks have a very complicated history. The surface of the moon is stripped of any nickel and probably one of the most important arguments against making the moon out of the same material as the earth.

SHOEMAKER: Terrestrial rocks have a couple of hundred ppm nickel. The difference is 2 orders of magnitude.

WASSERBURG: The earth has an iron nickel core. The best way to rip nickel out of anything is to dump it into metallic iron. It is very hard to oxidize. To make the moon–the best way is to have metallic iron in a box someplace, bleed the nickel which would be present in one of the silicates into the metal phase, throw the metal phase away. The earth, which presumably has a metal core, has a hell of a lot more nickel in the same rock types, so the lunar stuff looks as though it has been through at least two stripping processes of nickel.

MARGULIS: Where do you predict the nickel is?

WASSERBURG: You have to bleed nickel out of rocks and the best way is to use a metal phase. Did this happen in the solar nebula or on the moon? The distinction is clear between the earth and the moon.

ORGEL: Nickel goes into iron sulfide, does it?

WASSERBURG: No. Into iron.

ORGEL: So, the iron comes out of iron sulfide. I don't understand what that means, if it doesn't start as iron but comes out of an iron sulfide phase.

WASSERBURG: The cause of the confusion is that there are two arguments here. One is the presence of iron metal in the rocks, the absence of nickel in the iron metal. The second is there is no nickel any place. If there were nickel, it should have been in the iron metal if the metal existed as a separate phase. If it doesn't exist as a separate phase you may not use that argument. There is none in the rocks as a whole when you compare them with the equivalent of any terrestrial rock. No nickel, no nickel, wham; iron is here, so there is a stripping operation taking place on the material of the moon. Whether it happened on the moon or in the original condensation of the solar nebula is a rather critical question. The moon is the only planet with a low density—a funny planet.

ORGEL: How much nickel is there in the glass?

WASSERBURG: Up to a couple of hundred ppm.

ORGEL: Is it absolutely clear that that isn't the nickel you are looking for?

SHOEMAKER: Yes.

WASSERBURG: Glass is a separate issue. The glass has iron nickel spherules in it, obviously has been melted, and has every piece of evidence you would care to ask for for the presence of meteorite impact.

SHOEMAKER: It has iron nickel phosphate and carbide; all the meteoritic phases that are in the glass.

WASSERBURG: The rocks don't have any nickel material.

MARGULIS: Are you leading up to an iron nickel core for the moon?

WASSERBURG: No, I am talking about planets. I am trying to make it very clear that if you want to rip the moon off the earth—which I would like to do, in spite of what I apparently said on television—I don't remember what I said on television—it is a bad habit I have——

HOROWITZ: You had better get another script writer.

WASSERBURG: Or talk less. If you didn't have this nickel stripped so badly, you might be able to make a good case.

KAPLAN: Might this lack of nickel be just a localized condition?

WASSERBURG: Maybe. It may have happened in events preceding the moon's formation or within the moon itself. I want to remind you all of what you already know: The moon is a kooky place, very clearly distinguished from the earth. It has a very low density for a terrestrial planet. The original concept that the moon separated from the earth stemmed from the observation that the density was too low. In fact, a good way to make a low density planet is to take a planet like the earth with an iron nickel core and a relatively high mean density, and rip out the silicates from the outside and leave the high density component behind.

If we don't we must condense a separate planet with very little iron in it, it should have some nickel in it. If there is no nickel in it, it means how the hell did it get off the earth. I don't know.

Let's look at ages. The parents are uranium making lead 206, including products U235 making lead 207, thorium 232 making lead 208, K^{40} making argon 40; and rubidium 87 making strontium 87. These have half-lives long enough to be useful. It is difficult to determine the age of the individual rock fragments on the moon as well as to get a very peculiar number: The *age* of the lunar dust.

SHOEMAKER: You are not talking about the age of individual rock fragments as rock fragments, but the age of crystallization.

WASSERBURG: Yes. You could use this method to date the ashtray by which you would mean the last time since the ashtray had a certain state and what that state is will remain is to be seen. The date of the ashtray, in the case of the argon, couldn't be measured because it is too young.

That doesn't tell you how primitive or primordial anything is. The next slide will show the marks obtained in the presence of xenon 129 excess due to the presence of intermediate long-lived radioactivity, and if this is conceptually unclear somebody had better ask a question. At the start of the solar system, we were close enough to the time at which elements were synthesized so that radioactive elements with half-lives of 10^8 years and 10^7 years were present. After a few times 10^8 years, all these are dead and not seen. So, the mark of your being near the nuclear reactor is the presence of these two. I actually call this element Rx because its prescription is the element necessary. These are tracers.

RICH: Excuse me, you say you do not find these products?

WASSERBURG: Right, but on the moon, yes. First I will show you an object where we did find them.

RICH: Do you find them on the earth?

WASSERBURG: No, you can't tell that on the earth. It is too complicated.

RICH: I must have missed the point in the argument, not finding them implies that something is older than 10^8.

WASSERBURG: Yes. Would you ask the question again. I am sorry.

SHOEMAKER: I think it wasn't clear. These have been found, but in a meteorite.

WASSERBURG: I always assume that everyone reads my papers. My colleagues don't read my papers and I don't read their papers, but I figured this was a group of biologists and perhaps a more sociable and friendly lot. The age of the solar system is 4.6 x 10^9 years. It is the age of all meteorites but one and many meteorites show their radiation badge blackened. There are many objects in the solar system that are 4.6 billion years old that show that their badges were blackened but the moon is not one of them so far.

Here is a decay scheme–principally rubidium became strontium 87, the rubidium 87 atom decayed and the strontium 87 atom decayed. So we measure the relative enrichment of the strontium 87 relative to the strontium 86 and then we calculate the age.

* * * * *

EDITOR'S NOTE: Dr. Wasserburg continued to describe the method.

* * * * *

ORGEL: How many rocks are there so far?

WASSERBURG: That have been dated so far? I will cheat and I will tell you more or less every major rock sample.

HOROWITZ: But there are only two major rock types, you say.

SHOEMAKER: About a dozen crystalline rock fragments actually have had ages run on them.

WASSERBURG: I am saying that of the crystalline rocks, the things that are really identifiable–they were molten objects which cooled and crystallized and we are now finding the date since the materials were last melted. They turn out to be from 3.65–3.68 billion years. Every rock but one brought back from Apollo 11 comes out this way, these are older than any known rocks on the surface of the earth.

Here is another rock. At the risk of boring you, the data are extremely clear and self-consistent. They are reasonably precise and always come out the same, so you can barely force yourself into making more measurements.

Fragments from the dirt of the soil either plot where the low K rocks plot or plot up here [pointing] where a high K rock would plot, so the soil does, in fact, contain both components. The crystalline rocks from Apollo 11 are all 3.65 billion years old by rubidium/strontium, and for the potassium/argon ages they also give 3.7 billion years ages. The two methods are in agreement. The evidence is by four independent decay schemes that show that the rocks on the surface of the moon are younger–roughly by a billion years–than the formation of the moon as a planet.

MILLER: How much spread is there on the potassium/argon ages?

WASSERBURG: In the good data, 3.6 to 3.9, but you take the bad data and you get all sorts of crazy things.

Then there is a little peculiar rock, about 80 mg., that I called Luny Rock 1. It did not come from the same place as the other rocks. This rock has an age of 4.4 billion years and represents a fundamentally distinct environment from any of the other samples we see. Here is distinct evidence for areas on the surface of the moon getting biting close to 4.6 billion years.

ORÓ: Can you tell us the characteristics of this rock, petrographic, meteoritic, or otherwise?

WASSERBURG: Yes; you will find it very interesting. It is small so it requires microanalytical techniques. The rock is plagioclase, magnesium iron pyroxene rock with no calcium in the pyroxene, which makes it totally distinct from the other Apollo rocks, light in color. It has very little ilmenite, $FeTiO_3$ in it. It contains calcium phosphates in great abundance with fluorine and chlorine in great abundance–ranging up to almost 5 percent chlorine and fluorine in the phosphates.

MILLER: Is this the hydroxyl apatite?

WASSERBURG: It is a very fine grain rock, with two phosphates in it. It has fluorine and chlorine apatites and whitlockite in it, and all of these have 7 percent rare plus an interstitial glassy phase in which the potassium occurs in concentrations up to 15 percent.

ORÓ: It is low in titanium then?

WASSERBURG: Yes.

SCHOPF: The age is essentially a whole rock date? It is not possible to do mineral separates?

WASSERBURG: We could. We were pretty scared of moving on it. The whole sample is very different from anything we've seen. It was only 80 milligrams and the desire to dissect it is powerful. I am always outvoted in this by my colleague, Don Burnett, because it is a very valuable sample. The thought of using it up before we finish is rather frightening. There are two choices with the age: The age is 4.44 billion years or it is younger, like the others. If younger, its initial strontium is so great that it must have come from the most super-stupendous reservoir of rubidium on the moon. That doesn't sound so smart.

ORÓ: Have you concluded that it is a meteorite?

WASSERBURG: I can't answer that, that is a question of omniscience.

HOROWITZ: You can't exclude it. It could be a meteorite.

SHOEMAKER: It doesn't look anything like any known meteorite.

WASSERBURG: It doesn't look as if there are any possible ways that it can be a meteorite.

HOROWITZ: A unique meteorite.

ORGEL: It looks less like a meteorite than other lunar rocks?

WASSERBURG: Yes.

SHOEMAKER: It has petrographic affinities to things on the moon.

ORÓ: That is the kind of answer I wanted. What about the nickel?

WASSERBURG: We didn't run nickel. As I say, the sample weighs only 80 milligrams, it is extremely fine-grained.

ORÓ: It is a miracle what you have done.

WASSERBURG: All of the data were done on 2 grams and I am showing you probably less than a tenth of the data.

Here is the breccia, which is a mixture of fine grain soil and high K rocks.

Here is that embarrassing result which looks as though from grabbing a shovelful of fine dirt from the surface of the moon, we can tell the age of the moon. The lead/uranium/thorium work that was done by several workers and the lead/uranium/thorium ages for the lunar dust–excluding the coarse grain fragments in the dust–also agree at 4.6 billion years. By four totally independent measurements a bucket full of dirt that was shoveled up from the surface gives us an age which looks like the age of the moon.

SHOEMAKER: Gerry, I think you have to be fair and say these are model ages. We are assuming when we say *age* that this has been a closed system of some dimension.

WASSERBURG: That is right.

SHOEMAKER: That may be such an extreme simplification of this complex case that we may not know how to interpret a number like this.

WASSERBURG: Right, I agree.

These are only model ages but they are rather surprising. I don't advise you to go out in your garden and shovel a bucket full of earth and expect to measure the age of the earth as if you had ground up the whole earth. This is amazingly what it looks like on the moon–somewhat embarrassing but exciting.

* * * * *

EDITOR'S NOTE: Dr. Wasserburg went into greater detail as he continued to explain his data.

* * * * *

The gut feeling I have is that to explain this simply we must be looking at an onion skin of the moon which is really 4.6 billion years old, but which has had small melting, producing the crystalline rocks a billion years after the moon was formed. A source of the heat is the matter of some conjecture. The moon is hot, it clearly was made molten to some extent. If there is sufficient heat to produce these so-called lava flows by impact, then you don't have any problem. I call them *so-called* to distinguish between the maria and the lava flows since you have evidence in this mare to thoroughly clarify this issue. If it was by internal heating, the problem is how the moon ever cooled off. As Urey has pointed out, if the uranium and thorium concentrations of the moon are representative of the moon as a whole, then the whole moon must be molten, which it is not.

If you say that the scale height of the fill in the maria is a kilometer or two, then the outer 100 milometers of the moon must have been melted if you don't want the entire moon to have been melted. That is the cheapest price you pay.

If you drop the scale height, you drop the thickness and it becomes trivial. My prejudice now is to make it trivial. Basically, I think that the surface and most of the moon is pretty old, and that it has not been remelted very much. We are only looking at a relatively minor phenomenon which occurred at 3.65 billion years ago which, however, is responsible for many features. We are not looking at wholesale melting on a big scale. Apollo 12 will tell us a lot more answers to this.

RICH: Maybe I have missed something. With that picture, doesn't that suppose that the fines are not generated by this class of new rocks that were made after the moon was 1 billion years old, but that all the fines came from the early period and none from the later period?

WASSERBURG: The age of the people in this room is not calculated by asking everybody's age and multiplying by the mass, you see. This age measures a time *since*. If you ground up the earth today—which by our standards is zero years old—and you gave me a teaspoonful, I would measure the age and tell you 4.65×10^9 years.

MILLER: Is that with potassium/argon?

WASSERBURG: No, that is rubidium/strontium and lead/uranium/thorium.

MILLER: I thought you said the Apollo 11 site did have the highland material in it.

WASSERBURG: Yes, it has material mixed in it but it is surprising that it should be so damned well mixed that it represents an average for the whole bloody moon. That is a little embarrassing.

YOUNG: I thought from the point you just made that it should date older, unless you get a highland date.

WASSERBURG: This is hard to explain.

HOROWITZ: Gerry, is the following picture reasonable, that the moon was formed 4.6 billion years ago and it was during this early period that the basalt flows that compose the maria were formed?

WASSERBURG: No.

I don't know about the basalt flows, but the crystalline rocks, which are basaltic rocks, have been dated 3.65 billion years.

HOROWITZ: Are you offering any model, then, that will explain these observations?

WASSERBURG: I can offer two or three models. I would prefer to try that later.

Coffee Break

ORGANIC CHEMISTRY OF LUNAR SAMPLES

KAPLAN: I hate to start a lecture with an apology but after the performance of the two gentlemen from across town, and especially after you have been satiated with these elegant experiments of the Wasserburg Lunatic Asylum, I must.

The data I am going to present are results of work that was carried out partially at Ames Research Center by the NASA consortium and partly by us

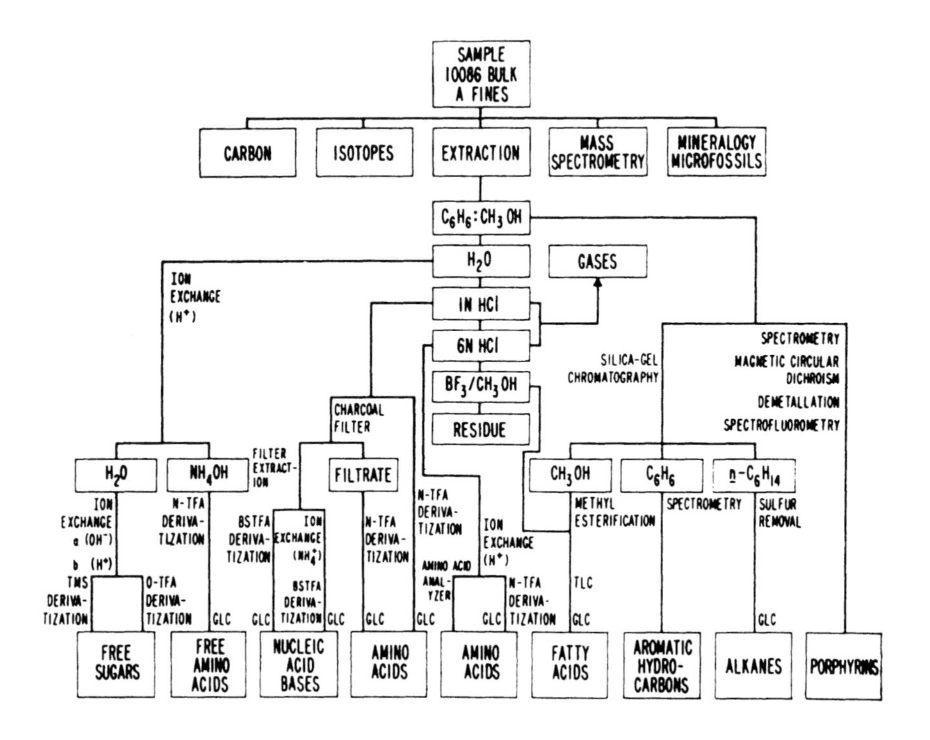

Figure 2. Scheme of analysis for organic components undertaken at Ames Research Center on Apollo 11, lunar material.

at UCLA. I want to describe the forms of carbon we believe we have recognized reliably, and give you some data on stable isotopes of carbon and sulfur.*

We began by attempting a very elaborate analysis for organic materials. At Ames and several other laboratories of organic geochemists we attempted to search for whatever we believed was important biologically, from experience with ancient sediments and meteorites. The sample was ground up, eluted, extracted with both organic solvents–benzene, methanol and inorganic solvents–water, acid–to look for hydrocarbons, sugars, nucleic acids, amino acids, fatty acids, aromatics, alkanes, and porphyrin. None of these compounds were found, with any confidence, in the samples that were assembled for the Ames consortium group, with the possible exception of a compound that has been labeled porphyrin.

However, this compound does not appear to behave like a typical terrestrial porphyrin. It had absorption and fluorescent peaks that were not identical to compounds separated from terrestrial rocks. However, it did appear to have very similar properties to a material that was extracted from some rock that had been exposed to lunar exhaust material. The conclusion that was arrived at by Hodgson and co-workers is that this material, having absorption and fluorescent characteristics of a porphyrin, may have been generated at the landing site of the LEM and, therefore, has nothing to do with organic molecules indigenous to the lunar material.

There is a possibility that some siloxanes, silicon-carbon-oxygen compounds were found. There was a group headed by Gerkhe who attempted to look for amino acids by carrying out an elution by hydrolysis and then making derivatives of these amino acids. They were unsuccessful, at least according to their interpretation, in identifying amino acids but they claimed that they did find peaks that came off the gas chromatograph. An effluent of the gas chromatograph was put into a mass spectrometer, and they claimed to identify–partially by analogy and partially from mass spectrophometric data–silicon-oxygen-carbon compounds.

The work is still in progress. The compounds they extracted may have been artifacts of experiments; they may have been dissolving rubber or silicon grease. But they were unable to manufacture them by actually introducing rubber and silicon compounds. These could not give them the hydrolysis product that they extracted from lunar material. Therefore they concluded that this material is indigenous to the lunar surface.

---

*All references given in this presentation can be found in the special issue of *Science,* vol. 167, No. 3918 (1970) and in *Geochimica et Cosmochimica Acta, Supplement 1, Proceedings of the Apollo 11 Lunar Science Conference* (1970).

I still put a large question mark there. I frankly am not convinced that this has been completely described with great certainty and I don't know whether Gerkhe, whose work is in progress, is completely convinced. I understand that Eglinton*, in England, also suspects that there may be some silicon-carbon-oxygen compound and that he is, in fact, working on some siloxane compounds which he believes may be indigenous to meteoritic material.

WASSERBURG: Has anyone calculated whether the concentrations are an order of magnitude correct to make this by siloxane production in the upper couple hundred of angstrom layers?

KAPLAN: The concentrations were equivalent to about 40–60 ppm of carbon that may be bound up in these siloxanes. This is equivalent to about 25 to 33 percent of the total carbon found–a very large proportion. It is somewhat difficult to believe that these siloxanes could really be major components.

ORGEL: This is very interesting. If you take the figures that are given for the amount of carbon in the solar wind, it comes out in this sort of range. By chance, I did it and I came out–I may have made an arithmetic error in this–with about 10 ppm for about a billion years.

ORÓ: Dealvo, in our group, made some calculations and they all hinge very much on what I referred to before as the sticking factor for carbon. Making certain reasonable assumptions he came to a concentration of about 200 parts per million, but I think Ian Kaplan was addressing himself at this moment in relation to these compounds of silicon. I doubt very much, based on our own work, if there are organic compounds in larger amounts than 1 ppm removable either by extraction or any other method.

KAPLAN: I agree.

ORÓ: So, I go along with your suspicion that it comes there from some other source. Further work has been done. Probably this material is a reaction product, or something.

KAPLAN: Let me summarize the results, since some of the people in the room may not be acquainted with other organic findings on the lunar material. B. Nagy's group in Tucson, Arizona, believes with confidence that a reasonably high content of organic molecules are present: complicated carbon-hydrogen compounds as well as possibly amino acids. Fox's group in Miami, Florida, found amino acids are present. And Eglinton's group, in Bristol, England, found methane was there in significant amounts as a carbon compound released in the gas phase.

Others suggest carbon monoxide and carbon dioxide are perhaps the dominant carbon compounds released from lunar material. I think Dr. Oró will discuss this. I now believe that carbon monoxide is probably the dominant gas released from the lunar material. Is this carbon monoxide

*Geoffrey Eglinton, Department of Chemistry, University of Bristol, Bristol, England.

released from lunar material present as gaseous carbon monoxide, or has it been formed as an artifact in the lab, perhaps by the water-gas reaction that Gerry Wasserburg mentioned earlier?

HOROWITZ: Where does the water come from?

KAPLAN: It may be partly terrestrial water. Quite a large amount of water was found by some. Freidman of the U.S. Geological Survey, for example, found water that may be terrestrial water.

HOROWITZ: Contaminant water?

KAPLAN: Yes.

WASSERBURG: Don't forget that the samples were never really kept dry.

HOROWITZ: When you say "released," you mean released by pyrolysis, by melting the rocks?

KAPLAN: There may have been two sources of carbon, non-volatile carbon and CO. Carbon monoxide may have formed either in the lab by the above reaction due to contaminating terrestrial water or at the lunar surface at some stage when both carbon and water were present there in small amounts. It may also have been present by reaction of non-volatile carbon with iron oxides or silicates to form CO plus Fe metal, both in the lunar surface and in the lab.

Calvin's group, at University of California at Berkeley, addressed themselves to this. They took some basalt I had sent him from Hawaii and they mixed it with carbon but could not form carbon monoxide. We repeated the same experiment–not with terrestrial basalt but with 100 mg. of lunar material, mixed with graphite–and we found large amounts of carbon monoxide. We could generate very large quantities, in fact, 6 times the amount of carbon monoxide that we obtained from just heating and extracting the lunar material; so this is a possible artifact.

These observations are important not only for an assessment of the amount of carbon, but also to our whole discussion of possible origin and significance of carbon to the evolution of life.

MARGULIS: Do you think this difference between your results and Calvin's have to do with the lunar basalt *versus* the terrestrial?

KAPLAN: I don't precisely know the method he used. Clayton–at University of Chicago–who did a thesis in Epstein's lab at Cal Tech 12 or 15 years ago, used this method to get oxygen out of rocks by reduction with carbon. It is strange that Calvin could not produce carbon monoxide this way, since this is now standard lab technique.

WASSERBURG: Lack of care is probably the answer.

KAPLAN: What we believe, in summary, is that we have evidence for iron carbides and possibly elemental carbon. We also think there may be some carbon dioxide. I will come back to this in just a minute.

WASSERBURG: Where are the iron carbides now, in the dust or in the crystalline material?

KAPLAN: In the dust. Most data I have are for the dust. The reason is that the amount of carbon in basaltic rocks is so small that we have no confidence at present that these very fine-grain rocks have any carbon. They come right at the threshold of the lab technique. The sloppiness of our technique is in the range of 10 to 20 ppm carbon. From igniting and reburning a sample, handling it, sieving it, we get some variation; one value on a blank was 15 ppm, another was 8 ppm. In the basaltic rocks we measured about 20 ppm carbon in one sample and 60 in another. If the basaltic rocks have any carbon at all, they have very small amounts.

Carbon monoxide and carbon dioxide were released by pyrolysis.

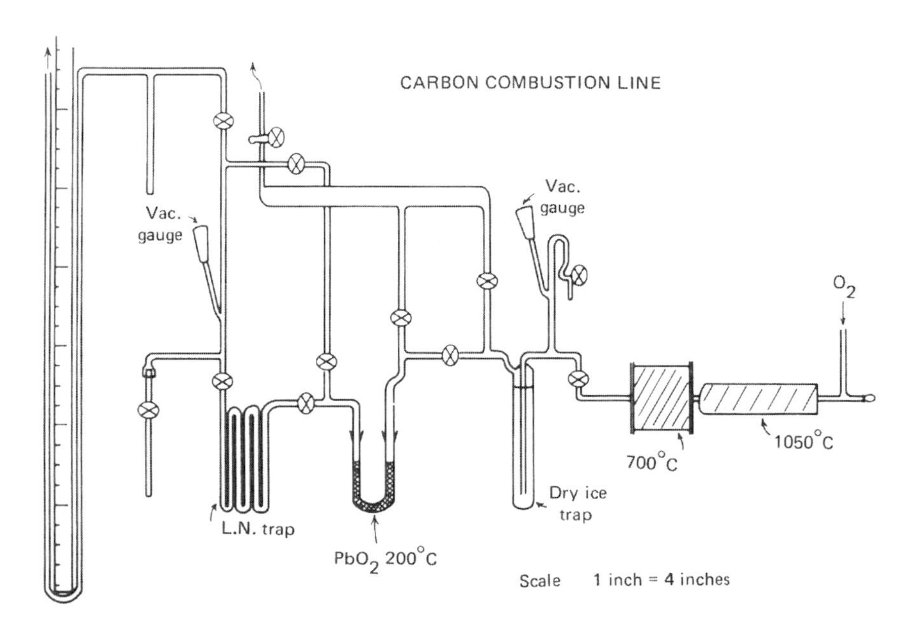

Figure 3. Combustion line for carbon analysis of lunar material.

This was done in an evacuated combustion chamber, and the carbon monoxide and other gases released were captured. An attempt was then made to identify them by gas chromatography and mass spectroscopy. In one experiment CO_2

was definitely identified, but in a repeat pyrolysis experiment on the lunar material, no carbon dioxide could be detected. Only CO was detected at temperatures of release from 500° to about 750°. Below about 500° very little carbon was released.

ORGEL: Are you going to tell us how much?

KAPLAN: Ultimately, most carbon comes off by heating. Several experiments were carried out. In one case heating from 150° to 250° gave 27 ppm gas; from 250° to 500°–these are two sets of experiments–19 ppm, 500° to 750°, 40 ppm; then taking the material that had been heated to 750° and combusting it, I obtained 59 ppm carbon in the residue.

The same series of heating temperatures was undertaken in another set of experiments: 150°, 250°, 255°, 500°, and 750°, and in this second sample we obtained about 130 ppm carbon. Then pyrolyzing the residue obtained after heating to 750°, by raising the temperature to 1,050°, 160

TABLE 1. $\delta^{13}C$ OF CARBON FRACTIONS RESULTING FROM PYROLYSIS AND HYDROLYSIS

EXPERIMENT		Sample Reference Number	Carbon Content (ppm)	$\delta^{13}C$ Per mil
Combustion of Intact Sample		10086	157*	+18.7*
		10084	132*	+19.1*
Combustion of Residue from HC1/HF Treatment		10086	30	– 7.9
		10086	42	– 4.3
Progressive Pyrolysis	150–250° C	10084	27	–25.1
Followed by Combustion	250–500° C	10084	19	–12.0
Of Volatile Products	500–750° C	10084	40	+23.4
and Final Residue	RESIDUE	10084	59	+ 9.4

** Average of values shown in Table 2.*

more ppm of CO_2 was obtained. In an attempt to combust the residue in an atmosphere of oxygen less than 1 ppm carbon was obtained, indicating that pyrolysis to 1,050°C virtually liberates all the carbon.

WASSERBURG: This is without the addition of any other gas?

KAPLAN: Yes. Was this liberated carbon originally gas or could it have been a nonvolatile form that reacted at elevated temperature and then came off as volatile carbon? At the moment, I don't believe we can say we have evidence.

By hydrolyzing the lunar material, we obtained a series of low molecular weight hydrocarbons: methane, ethane, pentane, butane. From a piece of meteoritic cohenite the same series of hydrocarbons were liberated in an identical experiment. Virtually no hydrocarbons were found in the blank controls. We reacted the lunar material with phosphoric or hydrochloric acid, and then measured the captured gases. No CO_2 was detected, indicating that certainly there is no carbonate–which nobody would expect–but also apparently no readily volatilizable CO_2 is captured in rocks that can be liberated by acid treatment.

We conclude that volatilizable carbon is largely in the form of CO, and we have estimated that iron carbides are present in quantities up to 20-50 ppm.

What is the evidence for elemental carbon? After we hydrolyzed the material to remove the iron carbide, the residue was reacted with hydrofluoric acid to remove the mineralized and crystalline silicate material, to leave a volume of only a few percent of the initial weight of the lunar sample. We finally obtained about 20 to 30 ppm carbon in the remaining material that reacted neither with hydrochloric acid or hydrofluoric acid. This we considered graphitic, or at least non-volatile carbon. Incidentally, graphite was found by Arrhenius. He identified one fragment; about a cubic millimeter of graphite.

MILLER: Was this in a thin section, or in dust?

KAPLAN: In dust.

MILLER: Are you sure it is lunar graphite?

KAPLAN: He claims it is hard to believe the graphite is a terrestrial LRL contaminant, but he is not positive.

WASSERBURG: I hope he didn't say that.

KAPLAN: What do you mean?

WASSERBURG: What got into the sample in handling? There are lots of complications. You may get a single grain of talc from talcum powder used when people put on their gloves—if there is a puncture in the glove you get contamination. I have that indelibly engraved in my mind. The presence of free carbon is a very pressing issue but it remains to be substantiated. We need cleancut, clever experiments to find out what the facts are.

MILLER: Are there any carbon isotopic data in the graphite that Arrhenius found?

KAPLAN: No, he just found a very small amount and doesn't want to part with it. I think Arrhenius believes the graphite is real.

In summary, I believe the important conclusion is that there is very little evidence for carbon hydrogen compounds. The lunar surface carbon is either largely in the form of carbon monoxide, perhaps carbon dioxide, or graphitic iron carbide material. The peculiar anomaly, if both carbon and hydrogen come down from the solar wind, is that we don't have methane, at least, in larger amounts.

TABLE 2. CONCENTRATION AND ISOTOPIC COMPOSITION OF CARBON AND SULFUR

SAMPLE		CARBON		SULFUR	
Description	Reference Number	Conc. ppm.	$\delta^{13}C$* per mil.	Conc. ppm	$\delta^{34}S$† per mil.
Bulk Fines	10086	143	+20.2	680	+5.4
		170	+17.2	640‡	+8.2‡
Bulk Fines	10084	147	+18.8	770	+4.7
		116	+19.5		+4.4
Breccia	10002,54	198	+ 8.4	1070	+3.5
		181	+ 9.2		+3.4
Breccia	10060,22	137	+ 1.6	1120	+3.6
		132	+ 2.7		+3.3
Fine-grained rock	10049	63	−18.8	2200	+1.2
		77	−21.4		+1.3
Fine-grained rock	10057,40	21	−25.6	2280	+1.2
		11	−29.8		+1.2

* *Relative to PDB standard*

† *Relative to Canyon Diablo standard*

‡ *Total sulfur by aqua regia oxidation*

We also undertook studies on carbon and sulfur isotopes. Table 2 shows a summary of what we find. On Earth heavy carbon is usually in marine carbonates. We proceed from heavy to light through atmospheric carbon dioxide, marine algae, generally in the range of about -15 or -20 ‰, marine invertebrates, coal, and ultimately petroleum. We have nothing much heavier than +5 on earth. Most of our terrestrial contaminants, things you touch in the lab, are in the range of about -20 to -30‰.

This shows both sulfur and carbon. The carbonate in meteorites is extremely heavy (+40 to 70‰). Meteorite carbon-hydrogen compounds tend to fall in a range from -15 to -17‰. From our recent studies we note an extremely narrow range for the carbon-hydrogen compounds in meteorites. Meteoritic graphite and carbide tend to range from -4 to -8‰. (Fig. 4.)

Now look at the lunar material (see Table 1, p. 68). The carbon isotope fractionation of the dust has a range of +21 to +17‰. The breccia range from about +10 to about +1‰, and the fine-grain basaltic materials from -20 to -30‰. We have no confidence in the basaltic carbon values because the amount was so low we were unable to get a good reading on the mass spectrometer—we had only enough for one sweep. The data could very well

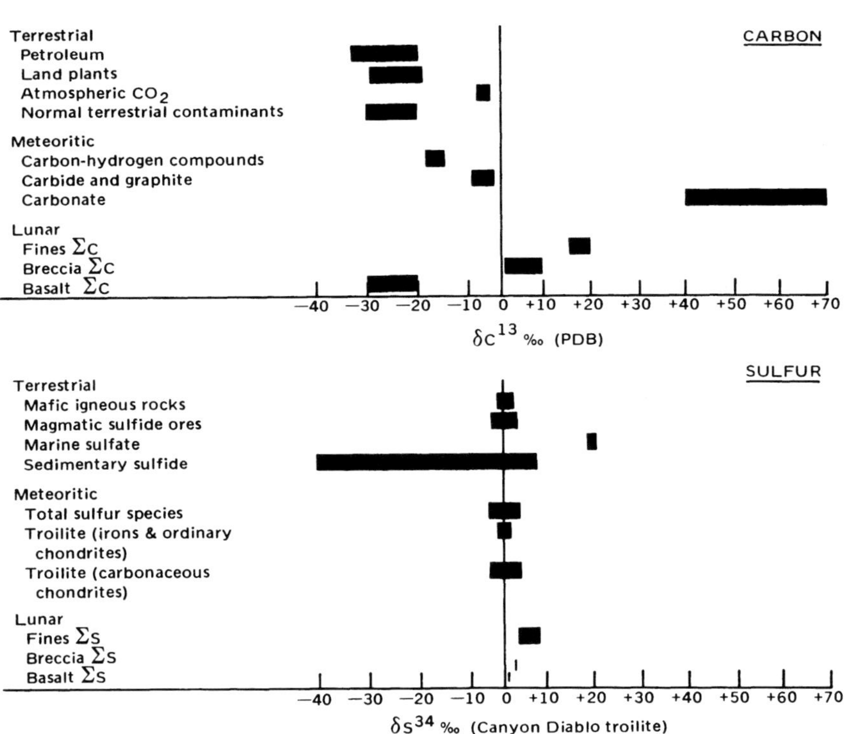

Figure 4. Distribution of δC^{13} and δS^{34} in terrestrial and lunar samples.

represent terrestrial contamination, and presently little can be said about the nature of the carbon in the rocks. The data on the fines and breccia are the most reliable results.

Most anomalous is the very heavy carbon in the bulk fines compared to terrestrial carbon.

Looking at sulfur, we see a rather similar situation.

WASSERBURG: How much total sulfur do you see in basaltic rocks?

KAPLAN: Total sulfur is about 2,000 ppm in basaltic rocks; about half that, about 1,200 to 1,000 in the breccia; and about 600-800, in the bulk fines. Zero is usually taken as the value for meteorites. Sulfides and sulfates in sea water are isotopically influenced by biological fractionation; the enrichment of the heavy isotope in sulfate is a result of preferential removal of the light isotope into sulfide during biogenic sulfate reduction.

Now, we see that the lunar dust falls outside the value for meteoritic sulfur; the carbon falls outside the range for all terrestrial and most meteoritic carbon, with the exception of the minute amounts of carbon from the fine grain basalt which may have been caused by contamination.

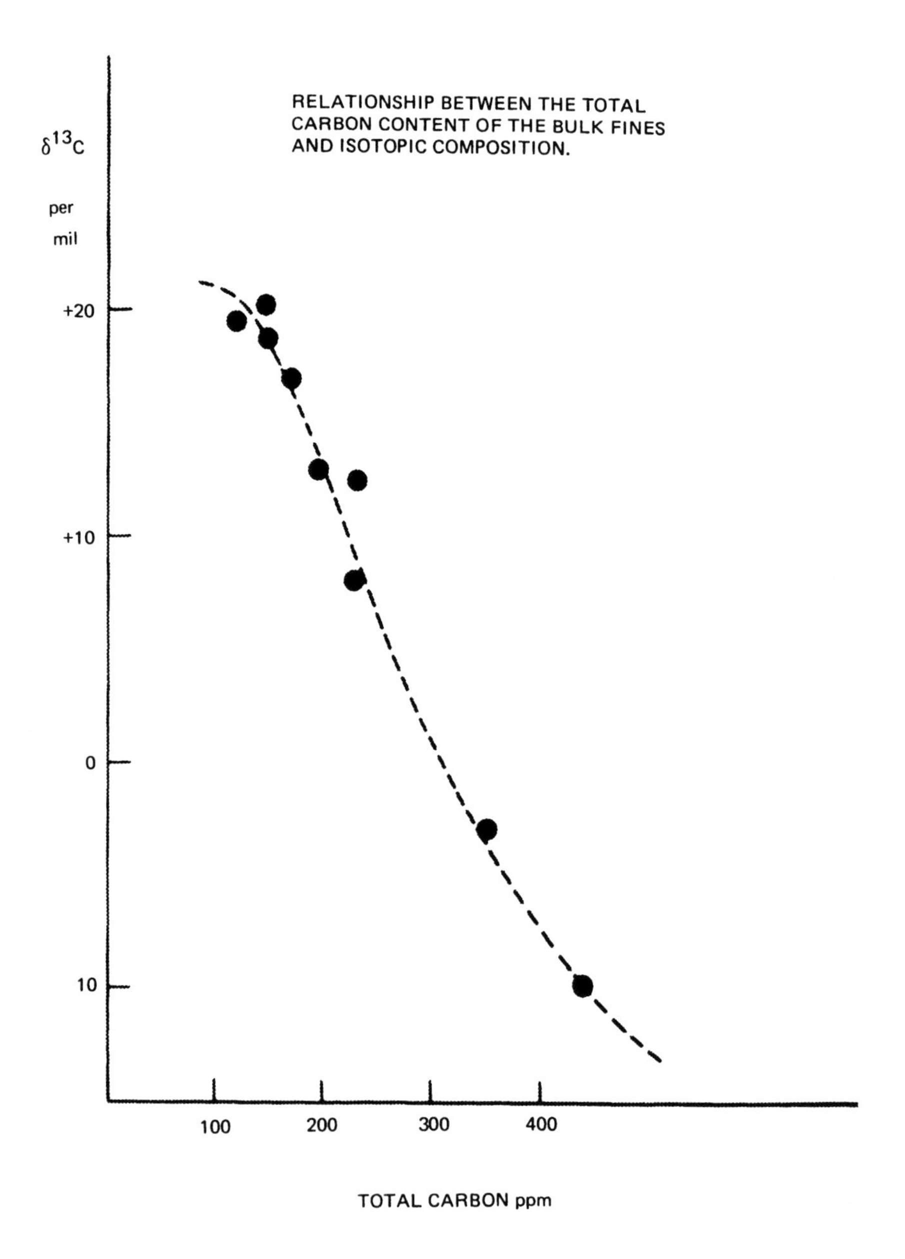

Figure 5. Relationship between the total carbon content of the bulk fines and isotopic composition.

Can we use this fractionation data to determine how much lunar fine-grain material has arisen from meteorites? I am arguing that from our isotope fractionation data we have no evidence that any of the lunar dust has arisen from meteorites during meteoritic impact.

Could we now identify how much of the carbon in the lunar material was contaminated by terrestrial carbon and how much was original? We found that the sample of bulk fines, containing the lowest carbon concentrations are isotopically the heaviest, whereas those displaying highest carbon concentrations were isotopically the least enriched–C^{13} from -10 to -12‰. This suggests that values much above 150 or 160 ppm carbon, are probably the result of a mixing of lunar carbon with isotopically lighter terrestrial carbon. A plot of the data falls on a straight line, and the only considered real values are those on the upper left portion of the curve (Fig. 5).

WASSERBURG: I am sorry, I don't follow you. The total carbon?

KAPLAN: Increases and becomes isotopically lighter. The higher the carbon content, the lighter it is.

MARGULIS: You are assuming the stuff on the right is contamination?

KAPLAN: Yes.

WASSERBURG: I still don't understand.

ORÓ: Because that is a measure of the sloppiness of the experiment.

WASSERBURG: Does this mean that when you have large amounts of carbon you have more contamination?

KAPLAN: Right.

WASSERBURG: But that is the reverse of what is expected. You should have had the least contamination when you had the most amount of carbon.

ORGEL: The high carbon is mainly contamination.

WASSERBURG: I understand, if you have a blank level, but the contamination should be most effective when you have the smallest amount. If you have a lot of stuff, then how the hell do you contaminate it?

ORÓ: You are right.

ORGEL: He is suggesting the amount of contamination is variable and whenever you see 400 ppm, that suggests–

WASSERBURG: That's just a poker game. In fact, if you said, "Go to the board, Wasserburg, and draw the expected correlation," it would be the reverse.

RICH: No, if you are a pessimist and realize that what they are doing at LRL contaminates things at random, then the pessimist would write a figure like that and say, "That is a reflection of the inadequacy of the techniques." Yours is the optimist's curve.

LEOVY: Do you call the left-hand end of the curve the real left-hand end?

SHOEMAKER: What was the size of the sample?

KAPLAN: The analysis here was carried out with 1 gram of sample. Most of the values yielding high carbon content and isotopically light were measured on a few hundred milligrams.

WASSERBURG: Now I understand it.

HOROWITZ: I think the answer to Conway Leovy's question is that there aren't any or very few terrestrial carbons with a +20.

KAPLAN: This is important because we are getting some spread in results on total carbon analyses and we must decide which results are reliable and which are not. Are the fines heterogeneous? Are the discrepancies we are now starting to find in the carbon analytic data due to heterogeneity or analytic techniques? We are not used to working with the material having such a high surface area and such low carbon content, and the carbon isotope fractionation data appears to be a good indicator for terrestrial contamination.

ORÓ: Actually, the method does not permit you to use the samples with a carbon content smaller than 200 parts per million to begin with. We calculated we needed at least 2 grams to make a meaningful analysis.

KAPLAN: Right. We were a little sloppy. We didn't know what to expect at the beginning and we learned the hard way.

It is of interest that everybody had found greater concentrations of carbon in the breccia than in the fines. But the carbon in the breccias was isotopically lighter. This at first could be explained on the basis of terrestrial contamination. However, I think it is difficult to explain it on the basis of terrestrial contamination, because every laboratory that analyzed these obtained approximately the same results. Second, if they were really contaminated like the fine-grain rocks, they would have been isotopically much lighter, which they were not.

Let's have a look at the sulfur. The sulfur had the highest concentration in the rocks–2,280, 2,200 ppm, with an isotopic ratio very similar–just a little heavier–to meteoritic sulfur (see Table 2, p. 70). The fines had the smallest concentration of sulfur and were considerably heavier than meteoritic sulfur. The two breccias samples were again intermediate, both to the fines and to the rocks.

In another experiment (see Table 1, p. 68) we attempted to determine the different carbons that are generated by heating. As the tempera-

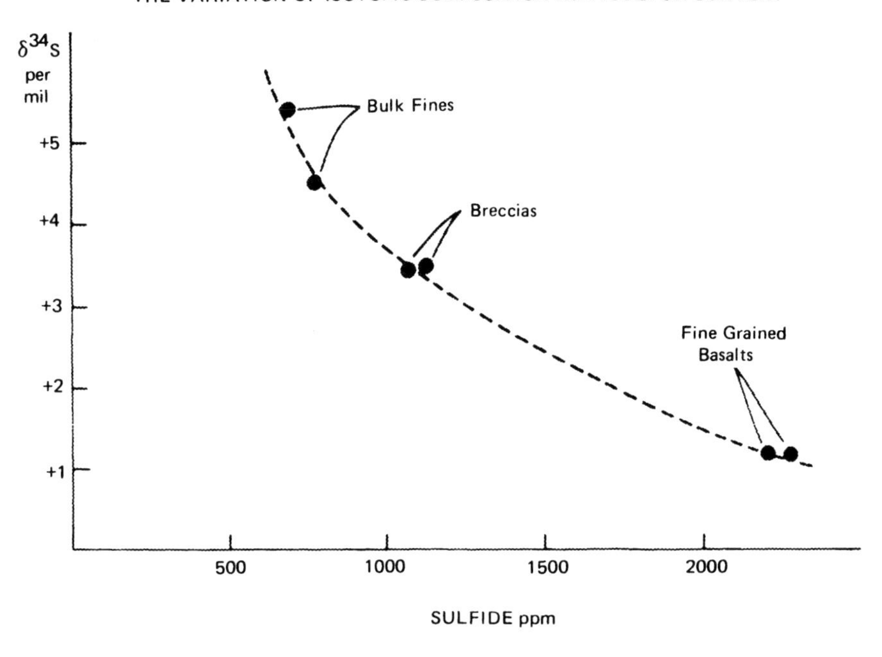

Figure 6. Variation of δS^{34} with total sulfur content of lunar material.

ture is raised, the amount of carbon increases, and the carbon released is isotopically heavier. We have found a very similar trend in another experiment since. However, again, we have no confidence that the carbon released during low temperature heating is necessarily indigenous carbon. It may have been a lab artifact. We captured only small amounts of gas. For isotope measurements we would like to have a minimum of 0.3 of a milliliter carbon. Here we are in the range of micromol quantities, extremely small amounts, barely sufficient to do an isotope analysis. It is possible that in some cases 30-40 percent of the products may have resulted from contamination during the extraction technique which involves a series of steps. We actually were able to generate very small volumes of carbon in the blank. We are confident, however, that at higher pyrolysis temperatures, where larger amounts of gas were obtained, the results are probably real. It is interesting that both from this and from a subsequent experiment, the material that was released at a very high temperature had a lighter isotopic value than the material that was released between 500 and 750°C.

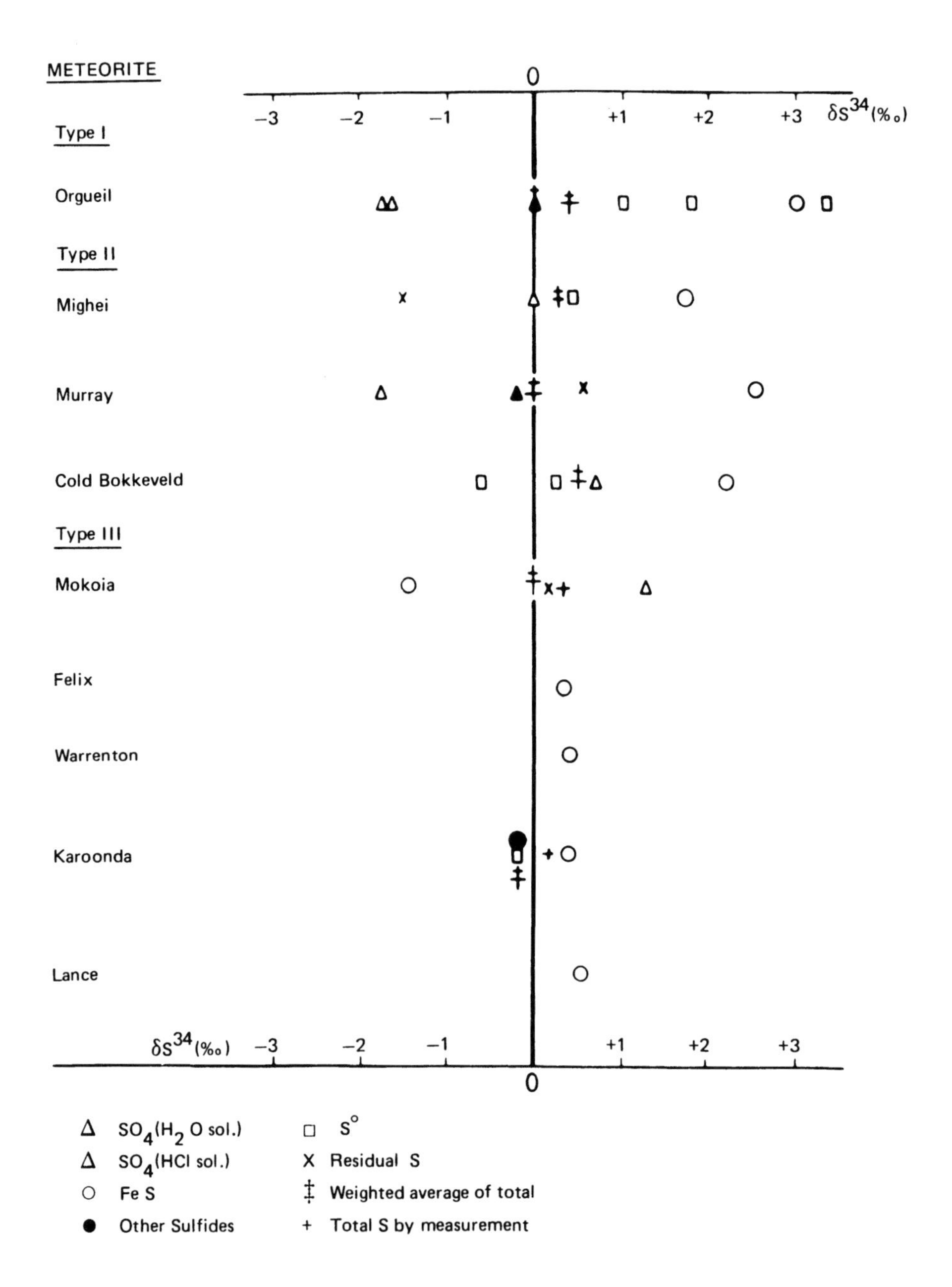

Figure 7. Distribution of δS^{34} in nine carbonaceous chondrite meteorites.

The carbon left after we hydrolyzed the sample with hydrochloric acid had an isotopic ratio of about zero or slightly lighter: -4‰. Again it appears that there may be different forms of carbon: One form is released at intermediate temperatures with a heavy isotopic value of +20‰ or thereabouts, and another, with considerably lighter isotopic ratios, similar to that found in meteorites, is released at higher temperatures.

ORÓ: What size were the samples in your hydrochloric acid treatment?

KAPLAN: We used 1 gram samples.

Figure 7 shows isotopic data of sulfur in meteorites and, as you see, the results are very similar, most within ± 2 per mil. For total sulfur in nine meteorites, the values all fall well within ± 1‰. The lunar bulk fines have values which fall outside the meteorite range thus arguing against a simple origin for any troilites from meteorites.

The carbonaceous chondrites which have sulfates, sulfides and elementary sulfur, have a larger spread for sulfur isotopes. But even with their large variety of sulfur compounds there is no spread beyond about ± 3‰, whereas in the lunar material the only sulfur compound of any significant abundance was iron sulfide and yet the total variation was ±7‰.

Figure 4 (page 71) shows isotopic data on meteoritic carbon. The total meteoritic samples fall in a range here from about -6 to about -30‰. The fact that the fines are generally about +20‰ argues again that there isn't a very simple mixture of meteoritic material with the lunar material to give an isotopic value of +20. From these studies, there appears to be no very simple contribution of carbon or sulfur by meteoritic material to the lunar surface. As Wasserburg said, perhaps our evidence is bad, but from the carbon and sulfur studies we have undertaken, we find no evidence for contribution of meteorites to the lunar material.

WASSERBURG: My comment was about the sulfur. The carbon is quite an impressive anomaly.

You hardly find anything with the sulfur but with the carbon it is a pretty clear effect.

SHOEMAKER: The difficulty is that there is much indigenous sulfur in lunar rock.

KAPLAN: The anomaly is the following: In the rock, there is an enrichment of sulfur and a depletion of carbon relative to the dust. What mechanism has enriched carbon in the dust relative to the rock, and also has enriched the dust in the heavy isotopes relative to the rocks?

SHOEMAKER: Can't there be fractional losses of sulfur by volatilization?

KAPLAN: Why shouldn't you have the same for carbon?

ORÓ: I don't think you can correlate these. Carbon, in the crystalline rocks is in very small amounts and the probability of contamination is high. This does not apply to the sulfur where the content is much higher. I don't think you can try to correlate these.

At the present time I don't trust any measurements on the isotope values of the carbon in the crystalline rocks.

WASSERBURG: But in the fines, where the money really rides, there is a problem.

KAPLAN: In attempting to solve this, two or three ideas have been put forward. One is that the isotopically heavy carbon has been donated by the solar wind. Since we don't know the carbon isotopic ratio in the solar wind, it is difficult to know if this is correct. Secondly, both the light sulfur and light carbon may have been lost by a melting process. Generally, high temperature processes do not tend to fractionate much so it is difficult to attribute this large fractionation to melting and escape of the light isotope. We put forward a third suggestion: The protons of the solar wind, rather than donating, may be stripping the heavy isotopes. Sulfur may be escaping as hydrogen sulfide and carbon may be escaping as methane by proton bombardment of carbon and sulfur in the lunar material. This might be causing an isotope effect in both these elements.

SINGER: This presents a problem. These compounds would be photodissociated before they would have a chance to escape. I think photodissociation would be much more rapid than the average escape time given by Jean's formula (Opik and Singer 1960).

KAPLAN: They would photodissociate into hydrogen and into carbon?

SINGER: Yes.

YOUNG: For the first of those mechanisms, why would there be greater enrichment in heavy carbons in lunar material than in meteoritic material, also exposed to the solar wind?

KAPLAN: The moon has a larger surface area of exposure than meteoritic material. In meteoritic material there is such a high amount of carbon that the small amount from the solar wind would be swamped and not necessarily recognized.

I would like to end on a note of speculation, since the lunar surface is ripe for that. We would like to suggest that perhaps the high concentration of carbon does not originate here at the locale where it is collected but it may have been coming from an area where carbon concentrations are higher than the now local area.

SHOEMAKER: Carbon or sulfur?

KAPLAN: Carbon. Carbon in the dust is higher than it is in the surrounding rocks or in the breccias. Perhaps it is continually coming from another source. A real problem is that the breccias have a higher carbon and if in fact breccias come from a welding of the lunar fines, carbon would be expected to have been lost.

WASSERBURG: It is very clear, really, that there is no evidence of differential volatilization for the dust or breccias compared to the rocks. If you look at the alkalis, for example, very volatile things like cesium relative to rubidium or potassium, the ratios of these elements are virtually indistinguishable between the rocks, the soil, and the breccias.

HOROWITZ: But carbon compounds are much more volatile.

KAPLAN: If it were a gas.

WASSERBURG: Maybe so, but in the process of making instant rocks, you talk about differential volatilization and I say it can't be done. Cesium is also pretty volatile.

SHOEMAKER: It is clear that you are not going to lose anything by going from regolith to instant rocks because they haven't gotten that heated. If they had it would clearly be recognized from the form of the rocks.

HOROWITZ: If the temperature rose to 1,500° for an instant–

SHOEMAKER: If it got that hot the associated shock pressure would be very easy to recognize in the specimen. They haven't been shocked very high.

WASSERBURG: At 1,500° cesium is as good a gas as sulfur.

KAPLAN: The lunar surface was supposed to melt at around 1,200.

HOROWITZ: I picked the wrong temperature.

KAPLAN: But if it existed as a gas, the shattering may have been sufficient to allow for its escape.

SHOEMAKER: If little gas bubbles had been broken, gas might have been lost, yes.

KAPLAN: In attempting to explain carbon and sulfur distribution, is it possible there is a mixing of the soil profile, and we are getting a density distribution, and the heavier sulfur may in fact be settling out here. The little glass spheres have a density of probably less than 3, whereas the iron sulfides have densities probably greater than 5, and we may have some stratification. May not the low sulfur in the regolith be due to the fact that there is some stratification differentiation and not necessarily a source origin differentiation?

WASSERBURG: But sulfur isn't lower in the soil than in the rock.

KAPLAN: Yes, it is. The sulfur content in the soil is about one third that in the rock.

MARGULIS: What are the generally accepted non-biological ways of fractionating isotopes?

KAPLAN: They can be differentiated by diffusion, simply following Graham's law: The rate constant is approximately proportional to one over the square root of the mass. For example, you could have a mass of 28 in CO composed of carbon 12 and oxygen 16, and a mass of 29 in CO containing carbon 13, so the fractionation will be the square root of 28 over the square root of 29. That is a substantial fractionation. It is possible that even during shattering there may be some diffusion differentiation through gas escape.

I don't know the real answers. These ideas are speculative because we don't have any real, hard data, only pieces of data.

But the problem for this audience is that the setting is just not there. In the absence of water, and of carbon-hydrogen compounds, the possibility of finding the processes needed for the onset of molecules important for life on the lunar terrain are very dismal. However, if there were a reasonable amount of iron carbides and some water, reactions might occur which generate compounds enabling us to have photoreaction, or whatever, to have polymerization. But if most carbon is in the form of elemental carbon or carbon monoxide, it does not look hopeful for carbon-hydrogen compound formation, and if Martian atmosphere or surface is in any way comparable to this situation, I think we are in for a nasty surprise.

MILLER: Distinction between Mars and the moon is that there is a lot of carbon on Mars, carbon dioxide in the atmosphere.

KAPLAN: But it is in an oxidized form.

MILLER: O.K., but very little has been found on the moon. Some people suggest you go down 10 or 50 meters, or whatever is necessary, and a great deal of carbon will be found.

WASSERBURG: Who said that?

MILLER: People speculate on this–

WASSERBURG: We have looked at stuff that is mixed to at least 5 meters.

MILLER: But the figure you usually give is 10 meters. Maybe the two will converge. There is no evidence, I realize, but is this at all reasonable?

SHOEMAKER: There is no basis whatever to think abundant carbon will be found below the surface. It is pure speculation in no way supported by what we have found.

MILLER: There is no reason for believing it is not there.

WASSERBURG: Yes, a fantastic number of particles have been looked at and even with low probability, were there a carbon mine somewhere–and graphite, for example, is very hard to destroy—it would be seen. I personally claim I have looked at 10^6 particles and I am sure there are others who have looked at more. Surely you would begin to see one little kooky thing.

RICH: One was seen. The question was, was it real?

WASSERBURG: So, if there were serious amounts of carbon after a certain amount of time we would begin to see some weird objects knocked out from deeper craters. I don't wish to be a negativist about it, but I think that holding the carrot in front of the ass all the time might tend to make you conclude there is an ass.

HOROWITZ: If there ever were organic carbon on the moon, you could explain why it is not there. It has all been pyrolyzed and diffused away.

KAPLAN: But there are real anomalies. The impact craters indicate logically that there must have been meteorite impact. I understand from statistical evaluations of observed falls that carbonaceous chondrites probably are the most abundant form of meteorites, at least landing on our own planet and, hence by analogy, on the moon. Where is that meteoritic carbon? It is difficult to assume rapid melting volatilized all that carbon and none was captured.

SHOEMAKER: The evidence suggests perhaps about 2 percent meteoritic material.

KAPLAN: Some evidence, not *the* evidence. It depends on which evidence you want to believe.

SHOEMAKER: Evidence from the lunar abundance, evidence of volatile elements that might have come from carbonaceous meteorites. An upper limit is 2 percent.

KAPLAN: Nickel content, perhaps.

SHOEMAKER: Suppose we take the typical carbon content of chondrites as one percent or a few: We may lose 90 percent of the impacting object for the high velocity particles; for the small craters this is generally true. Now we are down to a few tens of parts per million and that might easily escape recognition in solar wind presence.

KAPLAN: You are assuming a large loss.

SHOEMAKER: Almost certainly there is.

WASSERBURG: The big question, Gene, is whether or not there are small particles coming in from which there might not be a large loss.

SHOEMAKER: Objects slowed down by the Robertson effect so that they are falling in at escape velocity?

KAPLAN: Or shatter and don't melt.

SHOEMAKER: The plain truth is that the evidence we have indicates the bulk volume of this is very small compared to the larger problem.

KAPLAN: Wait a minute. If we assume that as Fred Singer says, there are photolysis effects, the meteorites land on the surface of the moon, the organic carbon becomes pyrolyzed, melted, and converted to either carbon monoxide, carbon dioxide, or methane. These are volatile materials that are expected to escape, assuming no pyrolysis occurs to cause polymerization into large compounds, and I don't think there are any logical arguments precluding such polymerizations. You assume that all these gases escape very readily and none besides a few tenths of a percent could be converted to elemental or any other form of nonvolatile carbon.

WASSERBURG: He is jetting them away.

SINGER: Could I add a caveat to my earlier statement? When we talked about thermal escape, I was referring, and I thought you were, too, to the escape at the present surface temperature of the moon. Even the highest surface temperature corresponding to the lunar moon gives a much slower rate of escape compared to the highest range for photodissociation, which is only about a day or so. With impact extremely high, local temperatures are reached and the rate of escape precludes inhibition. The rate of escape is very fast; this must be kept in mind.

Furthermore, small particles produce just as much of the high temperature as larger particles. We must not be misled by the experience on the earth where small particles are slowed down by the atmosphere. Small particles on the moon will produce on a smaller scale exactly the same physical effect as very large impacting meteorites.

KAPLAN: Assuming the whole meteorite virtually shatters and melts.

SHOEMAKER: That is right, unequivocally. Almost nothing typical of the high velocity material will survive melting. These little sheared metal grains, small in abundance, have just kind of fallen on the moon and they cannot have been at very high speed. But in these high-speed impacts, the bulk melts and a very significant fraction is volatilized right away.

KAPLAN: Therefore, no significant amount of carbon is expected to be captured by the lunar surface?

SHOEMAKER: I think the amount might be significantly smaller than the solar wind current so that it can't be recognized isotopically.

ORÓ: I think probably something like 10 ppm is meteoritic. By taking 1 percent theoretical composition, 1 percent carbon and then 10 percent remaining, the order is 5 or 10 ppm of carbon. That is not unreasonable; this contribution to carbon that might be lunar or solar would not significantly alter the isotopic conclusion.

I must admit that I, too, probably pre-judged and, in fact, I was influenced by Gene on this large meteoritic fall in Canada. I thought the carbonaceous chondrites falling on the moon might make the largest percentage of meteorites, leading to a higher percentage of carbon. I think the results now tell us that this may not be the case. I confess right here, too, that the results are inconsistent with hitting the moon, as someone has argued.

HOROWITZ: I thought it was agreed that if carbonaceous chondrites were hitting the moon, they would melt and the carbon compounds would diffuse away.

ORÓ: I assumed about 10 percent would be retained, which is the value that he gives.

SHOEMAKER: In a committee meeting a year ago or longer we guessed there would be 50 ppm carbon from carbonaceous meteorites in the fines, and as we find about 200 ppm total, we should recognize isotopically this contribution.

KAPLAN: How can we account for the depletion of sulfur in the fines relative to that in the crystalline rocks?

WASSERBURG: I am not sure, 5:1 ratio—

KAPLAN: 3:1, 4:1.

WASSERBURG: I think the sulfur argument is the weakest. Carbon is a problem, besides the effect, it is a very peculiar amount. This is real kooky carbon of extra-terrestrial origin and I don't understand it. The carbonaceous chondrites certainly are not excluded. If you want to play the record another way, I thought there were going to be a hell of a lot more pieces of iron on the moon, but I was wrong.

SHOEMAKER: Why didn't you tell me that? I would have bet you against the iron and won back my bottle of champagne.

WASSERBURG: We argued and I was sure. I expected not glass balls but a lot of iron balls but I didn't find them. The iron is there. In fact you get stupid phone calls. Some jerk will say, "It's funny the transparent glasses are ferromagnetic." Indeed they are because there is little iron in those spherules, but damned little.

KAPLAN: I find the escape of the carbon molecules difficult to understand: If the glass spheres cooled very rapidly—and they had to—why wasn't at least some carbon captured?

SHOEMAKER: Surely some must be.

KAPLAN: But we have no evidence for this whatsoever. Intuitively you say some should have been captured and I agree, but I say we have no evidence.

SHOEMAKER: We are looking at shock-melted stuff and it is clearly partly contaminated with meteoritic material and we get some estimate of that value. Let's suppose we recover 1 percent of residual meteoritic contamination, and typically the stuff hitting the moon is 10 percent carbon.

These are reasonable values and if it is mixed with something else like solar wind carbon, we don't know how to label and find it.

KAPLAN: This is the story that I wanted to give. We have data, which at the moment is difficult to interpret. Maybe Dr. Oró has another interpretation from his data. Lastly, as far as organic evolution, we don't at the moment from this analysis have a story to tell.

LEOVY: Wasn't there a statement made earlier today about carbon monoxide within the glass spherules?

ORÓ: Yes.

LEOVY: Is this going to be discussed?

SHOEMAKER: Bill Schopf, do you have to leave now?

SCHOPF: I think so, Gene. I don't have any great insight to offer to the group regarding the state of carbon or on the biology of the moon. Briefly, we spent a fair amount of time looking at lunar material, both at Houston and UCLA, and at the Ames Research Center. We found a surprisingly large amount of contamination: particulate organic matter, fibers, Kleenex, a wide variety of organic junk. I think that all analyses of Apollo 11 organic matter without some sort of handle, such as isotopic values, may be considered suspect.

Particulate organic contamination was detected at Houston during our preliminary studies, and in some samples distributed to at least some principal investigators, there was similar organic contamination. I can of course show you pictures of organic contamination, glass spheres, et cetera; these, however, would not be particularly instructive. Like Gerry and everyone else who has a scanning electron microscope, we studied these things because they are kind of fun.

WASSERBURG: I would like to make a statement while you are present. I think the biological and biochemical community, interested in the problem of organic matter, however primitive on planetary surfaces, is responsible or will have to become responsible for designing experiments which will work within the existing framework to actually test specific hypotheses. They must not depend on the whole confounded complex of NASA to deliver pristine samples to them because they don't exist, and it is hard to believe they could ever exist.

I have worked together with some very distinguished and conscientious members of the community, and we worry about these things. There is one answer: You guys–in a community sense–must formulate packages to bring back samples isolated enough under satisfactory environments for you to perform these very important experiments. If this cannot be done on moon samples, you ought to cancel the bloody Mars shot.

SCHOPF: Your remarks are well taken and it must be said that the people at the Manned Spacecraft Center did a very good job regarding organic contamination for a first time through.

The problem is in distinguishing between a sterile situation and an organic-free situation. They are not synonymous although some people seem to think they are.

The pre-cleaned can that was sent up with Apollo 12–and I gather that will be sent up with 13–to be filled with fines from the lunar surface was an attempt to get around this problem.

Another possibility would be to have individual investigators provide their own clean-sample containers with all the blanks run, but at present this would also be a terrific problem, since it would involve a "clearing-house" at Houston.

RICH: Since we are not bringing back samples, I don't see the relevance to Mars.

WASSERBURG: If you are concerned enough about the problem to run the Mars Program through, you have responsibility to be concerned about a somewhat closer control planet with several missions, like the moon, to try to attempt to get some intelligent information. The Mars extravaganza is even less justified if you have not run at least a small show first with enough concern to find out what the problems are.

RICH: But the problem with Mars is not that of getting back a sample without contamination. There we must do an analysis *in situ,* the problems are great, but they are different from the moon.

ORÓ: Gerry, we are very much concerned, and perhaps even more than you are. My experience tells me that a reasonably good if not wonderful job has been done. Many laboratories have had much worse contamination, probably higher by one or maybe even two orders of magnitude, than what they have now. It is not perfect, I admit it, but I am amazed at the wonderful job done. The Viking, as Alex said, presents a different problem. If it were possible we would remove all the plastic materials that go into it. This is impossible because the industrial revolution is based on the application of plastics, and so forth, so we design to minimize plastic.

RICH: One important area of overlap is the site contamination due to the retrorockets; although the lunar and Martian shots use different fuels, the problem of site contamination is important for both. Is it right that the lunar material had no detectable level of contamination?

ORÓ: I wouldn't say that. The results from the laboratories suggest the organic substances detected come from contamination.

HOROWITZ: Mostly carbon.

ORÓ: Yes, but realize that we are talking about levels which I think are of the order of 1 part per million. If we are going to find life on Mars, we are not going to make a decision on the basis of 1 part per million, because then it is a lost cause. If there is life it is going to be several parts per million.

RICH: Three.

MARGULIS: If the fuel tank exploded and dropped its organic compounds on the surface right at the site and, say, you looked at it 2 years later, what kind of distribution of these organic molecules would you see over the rest of the moon? Do you have any feeling for that? If porphyrin is found on the other side at another site 2 years later, could it have just been in this mixing layer?

ORÓ: That's a good one. There is a lot of organic stuff in the lunar fuel which is not being used for Mars. It is basically hydrazine, you eliminate the nitrogen, and then if you find significant carbon it still would be valuable, meaningful information.

SHOEMAKER: Part of the answer to your question is that the amount of gas-released combustion products in the descent of the LEM exceeds by a large factor the probable mass of the lunar atmosphere, if it has any. So with each landing a lunar atmosphere is created and almost certainly distributed around the moon very rapidly. In fact, the organic compounds are probably produced in the combustion rather than in the unburned hydrogen.

SINGER: I doubt this very much. You must think of the physical picture, of what happens to the organic molecule on the moon. It describes a ballastic orbit and its velocity will depend on the local temperature at the last point at which it hit, and the point at which it is emitted. On the day-side of the moon it hops because the temperature is high and when it gets over to the night-side of the moon it makes little hops.

SHOEMAKER: You are saying it comes down quite rapidly.

SINGER: It hops around until it freezes and it can freeze out at any point that is accessible to the sun, when the sun comes around or the moon turns 14 days later, this molecule then re-evaporates and continues hopping around.

The lifetime of such a molecule against photodissociation is a day or a fraction of a day, depending on the molecule. It becomes successively photolyzed, the very simple molecule equals an atom in the end, and terminal escape will be very rapid.

There is another caveat. There are lunar regions always protected from the sun. These regions are near the poles of the moon. Volatiles of all kinds are expected to accumulate there and form a glacial sedimentary type of stratographic record. I believe this is probably one of the most valuable things to examine on the moon. I would expect to find the stratographic indication of the lunar landings corresponding to the year 1969 in these glaciers as quasi-glacial deposits near the polar regions of the moon. I look at the artificial atmospheres we are producing now as convenient indicators.

SHOEMAKER: You are saying that they reach their lifetime, which indeed they do, and they precipitate on the moon surface rapidly and then the migration is by progressive hops.

SINGER: That is right.

SHOEMAKER: And you have a little less than 1 percent of the area near the poles where these will trap out. It turns out that almost all the water will go to these cold traps.

SINGER: No, because it can escape first.

SHOEMAKER: No, the chance of getting to the cold trap is high relative to escape.

ORÓ: I would like to use this opportunity to bring to everybody's attention that we exert any influence that we can to have one of the landings be at the poles. I don't know, Gerry, whether you can be effective in transmitting this information. I think a number of people agree—that we should look for organic material at the lunar poles.

WASSERBURG: I think you guys are doing a lousy job looking for it around the equator.

ORÓ: Why?

WASSERBURG: I think you have to design flying experiments where materials can be covered on the surface of the moon, safely canned, and brought back to the laboratory adequately shielded under extremely clean conditions so you can actually say what comes back from the moon. To talk about an extravaganza like polar landings compared with making up a double package of clean samples from an equatorial site is hard for me to find reasonable.

ORÓ: I am sorry, I think you are scared. As I said before: The bulk of Sample 82-16 in our laboratory has no measurable amount of organic extractable compounds. To be precise, it has less than 0.001 part per million.

WASSERBURG: It is funny though—when Schopf looks at it that he finds all sorts of hairs—somebody else finds quartz. There is a problem and I think there should be more critical biological material——

MARGULIS: But to look at it you have to put it on a slide on the microscope stage and just by doing that you can't help but get fibers in it. That is irrelevant. The fact of much less than a part per million extractable organic material is the best evidence there is that absolutely nothing of direct biological interest is there.

ORÓ: Gerry, you cannot distinguish what was put there by the investigator himself, so if you are saying that, that is all right.

HOROWITZ: Gene, including Singer, we will have two and maybe three speakers for important information on the moon. I would like to propose that we have a session after dinner to run for an hour or however long, in order to finish the moon today. Is that acceptable?

The conference recessed at 5:45 P.M.

Saturday Evening Session

The conference was called to order at 8:30 p.m. by Dr. Gerald Wasserburg.

WASSERBURG: May I call attention to the fact that we are in session. Since the acting chairman is gone I am the acting chairman and we will proceed. Professor Oró, you are on target.

ORÓ: I will try to make it brief. There are 21 slides so it shouldn't be more than 21 minutes.

Gerry Wasserburg's concern has been in our minds for a long time. It was most obvious when I saw some work presented on nucleic acids and meteorites in 1964 in San Diego, California. I concluded then that there is a point of diminishing returns in attempting to find traces of certain organic compounds in meteorites.

There may be something similar in the biological sciences to what we have in the physical sciences with the uncertainty principle: Life cannot detect life because it carries so many side products, whether they are living organisms or products of living organisms, that it is practically impossible to find an environment which is exceptional and will not influence the measurements. Therefore, we have to set up practical limits for the significant concentration values and even if over-simplifying it, I adopt the position that I should not spend much time thinking about the significance of anything organic that is present in the lunar samples in amounts lower than 1 ppm or 0.1 ppm. Actually we knew in advance, from the results of Turkevich on the Surveyor and now from the results on the Apollo samples that the analyses match reasonably well those of some igneous rocks, and in an igneous rock like a piece of lava we cannot expect to find much organic material.

I would like to talk to you about 10 analytical experiments we have performed on the lunar samples from Apollo 11. These are listed in Figure 8 and they are concerned with analysis of the organogenic elements and compounds that are necessary for the synthesis of organic molecules, and with related measurements of potential significance for an understanding of the problem of chemical organic evolution in the solar system. Most of the work has been done with samples of lunar fines. A very small amount of work was done with a breccia sample.

First, the organogenic and related elements were measured by spark source mass spectrometry. Second, vacuum differential thermal analysis and high resolution mass spectrometry were used to measure the gases evolved from 100° to 1400°C. Thus, among other gases, N_2 and CO—both with mass 28—were resolved and measured on a photoplate. Third, the gases evolved up to 750°C, including CO, CO_2 and traces of organic molecules, were measured by direct quadrupole mass spectrometry, using a minute low resolution instrument of potential application to the unmanned exploration of planets–e.g., Mars. Fourth, the gases that evolved every 100°C intervals from 100 to 750°C were first separated and then identified by means of a medium resolution LKB gas chromatograph-mass spectrometer. Fifth, the latter instrument was also used to separate and identify the volatile compounds generated upon acid treatment of the lunar fines. Sixth, the organic compounds evolved by progressive heating to 480°C–excluding CO, CO_2–were separated and detected by flame ionization gas chromatography. Seventh, the organic extractable organic compounds were analyzed by flame ionization gas chromatography. Eighth, any amino acids resulting from the 6*N* hydrolytic treatment of the lunar fines were derivatized and gas chromatographically analyzed. Ninth, the carbon-13/carbon-12 ratios of the bulk and sieved lunar fines were measured by means of an isotopic ratio mass spectrometer. Tenth, an effort was made to localize any solid form of carbon in the fines and breccia by means of a direct imaging ion probe mass spectrometer.

A more detailed description of these experiments and the techniques used may be found elsewhere (Oró et al. 1970a; Oró et al. 1970b; Oró et al. 1970c; Gibert and Oró 1970; Gibert et al. 1971; Updegrove et al. 1970; Socha et al. 1970; Wachi et al. 1970; Wachi et al. (in press); Updegrove and Oró. 1969). A brief presentation will be made in the following slides.

Table 3 shows the results on the composition of the organogenic and related elements. Let me point to some values. Realize that we are probably looking to values for elements in solid form. The upper value for carbon is 41. Values for residual carbon of 36 to 73 ppm have been reported by other investigators. The upper values for nitrogen are about 20, phosphorus about 270, for sulfur about 4,200.

WASSERBURG: What was the weight of the samples you used?

ORÓ: About 250 mg for the elemental analysis in the 1st experiment and about 35 milligrams for the following gas analysis in the second experiment. Figure 9 has curves for argon, helium, neon, CO and CO_2 evolution from 0 to 1400°C. Let me give you the conclusions in advance. The release of carbon dioxide is similar to that which could be expected from the decomposition of a carbonate. Calcium carbonate in the form of aragonite has been found in the lunar samples. However, some CO_2 may be absorbed from the terrestrial atmosphere. The carbon monoxide is either chemi-adsorbed, combined as carbonyl derivatives, or results from a chemical reaction. Figure 9 also shows the evolution of some noble gases for comparative purposes.

WASSERBURG: No hydrogen?

ORÓ: It was not measured in this experiment.

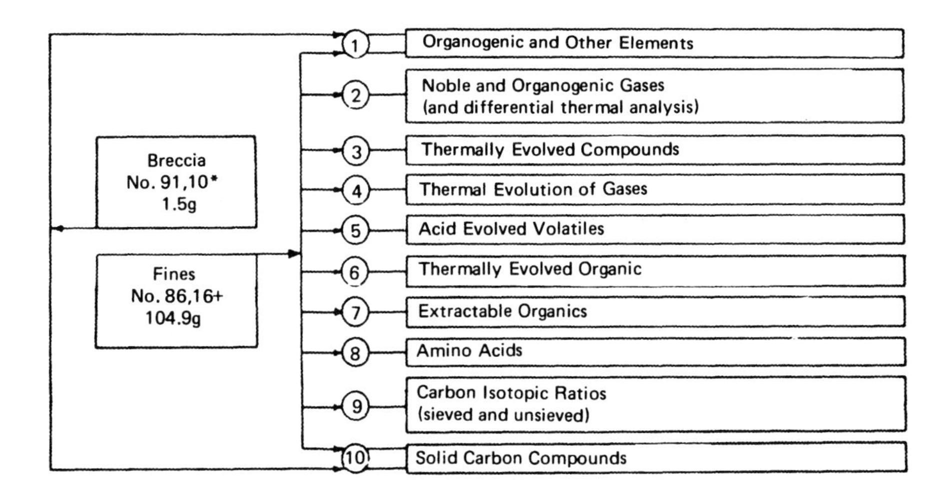

Figure 8.. Schematic flow diagram of sample distribution for the analysis of organogenic elements and compounds in lunar fines and breccia from Apollo 11.

TABLE 3. SPARK IONIZATION MASS SPECTROGRAPHIC ANALYSIS OF ORGANOGENIC AND OTHER ELEMENTS IN APOLLO 11 SAMPLES.

Element	Detection Limit	Rock		Fines	
(%)					
Fe	0.3	12	15	10.4	12
Ca	0.3	5.8	9.2	6.3	8.0
Mg	0.3	3.3	4.6	4.2	6.3
(ppm)					
S	5	1000	2800	590	4200
Na	0.07	340	2600	1300	2000
Mn	0.1	340	2600	1400	1650
K	0.07	1200	2300	600	3600
P	0.3	59	250	110	270
N	2	<10	15	7	20
Zn	5.0	N. D.	N. D.	<5	14
Cu	3.0	N. D.	N. D.	<3	12
C	1	5.5	6.0	5	41
Co	0.3	3.3	14	8.7	20
Cl	0.3	1.4	3.3	0.91	3.5
H	0.03	0.08	0.13	0.41	1.2

N.D. = not detected.

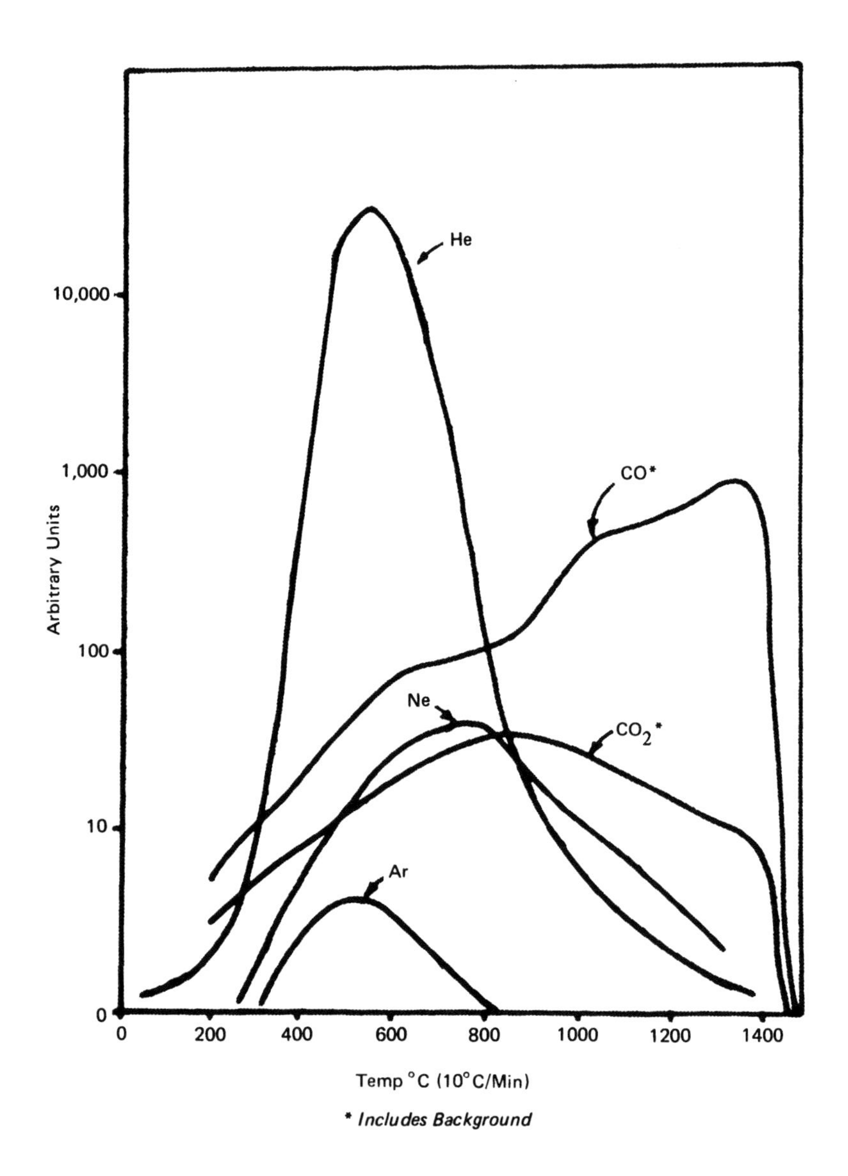

Figure 9. Gases released as a function of temperature in a high vacuum differential thermal analysis (DTA) cell.

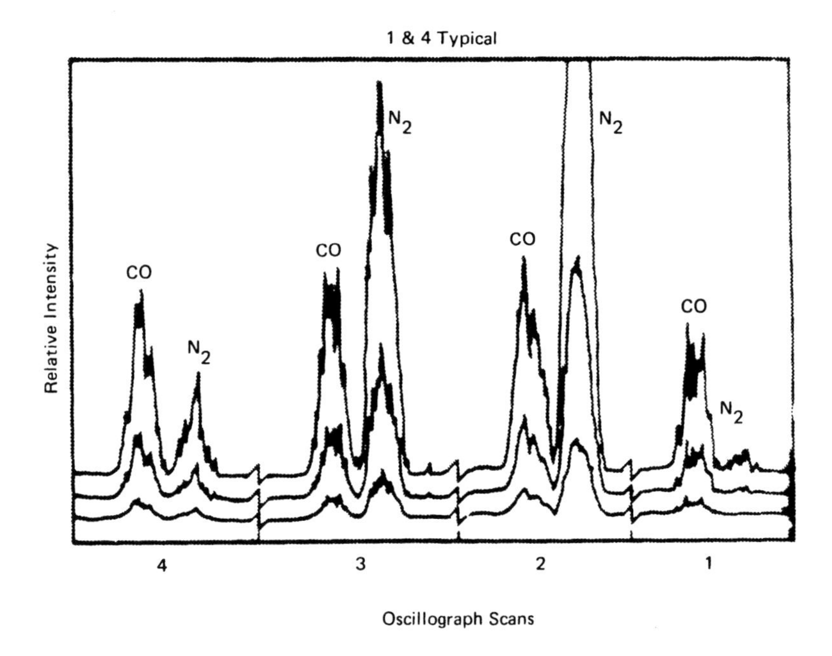

Figure 10. High resolution mass spectrometric ion intensity for CO and N_2 recorded every 2 seconds (at 914°C).

Figure 10 shows very sudden releases of gases, at a temperature of 914°C. The scan is taken every 2 seconds with the high resolution mass spectrometer. There is usually more CO than nitrogen but now and then a big burst occurs where a predominance of nitrogen is observed. Whether these are bubbles or not, I really don't know, but at the moment I don't have any other interpretation. Gerry, what is your interpretation?

WASSERBURG: The trouble is, with a spark source—

ORÓ: This has nothing to do with the spark source. This is a high vacuum DTA–differential thermal analysis–cell where the sample is heated from 100° to 1,400° and you scan continuously for the masses that you are interested in.

WASSERBURG: This sudden appearance is hard to explain. I would have a better explanation if you had done it on the spark source.

ORÓ: This second experiment has nothing to do with spark source.

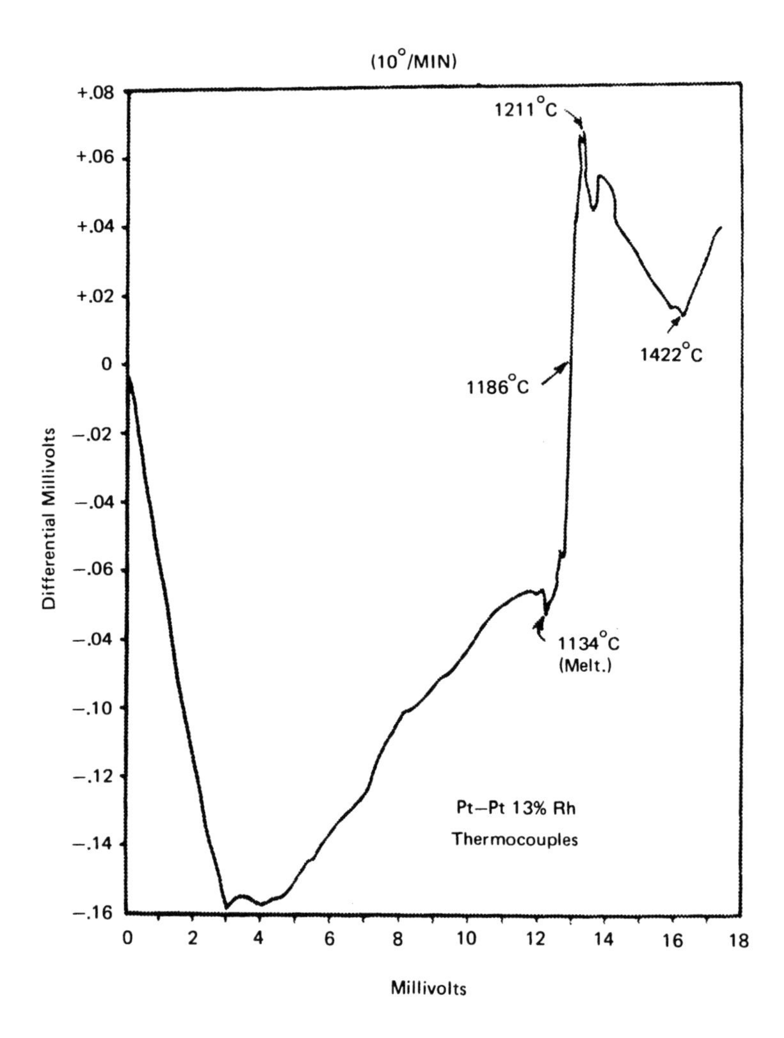

Figure 11. Differential thermogram for the temperature range 200-1500° C.

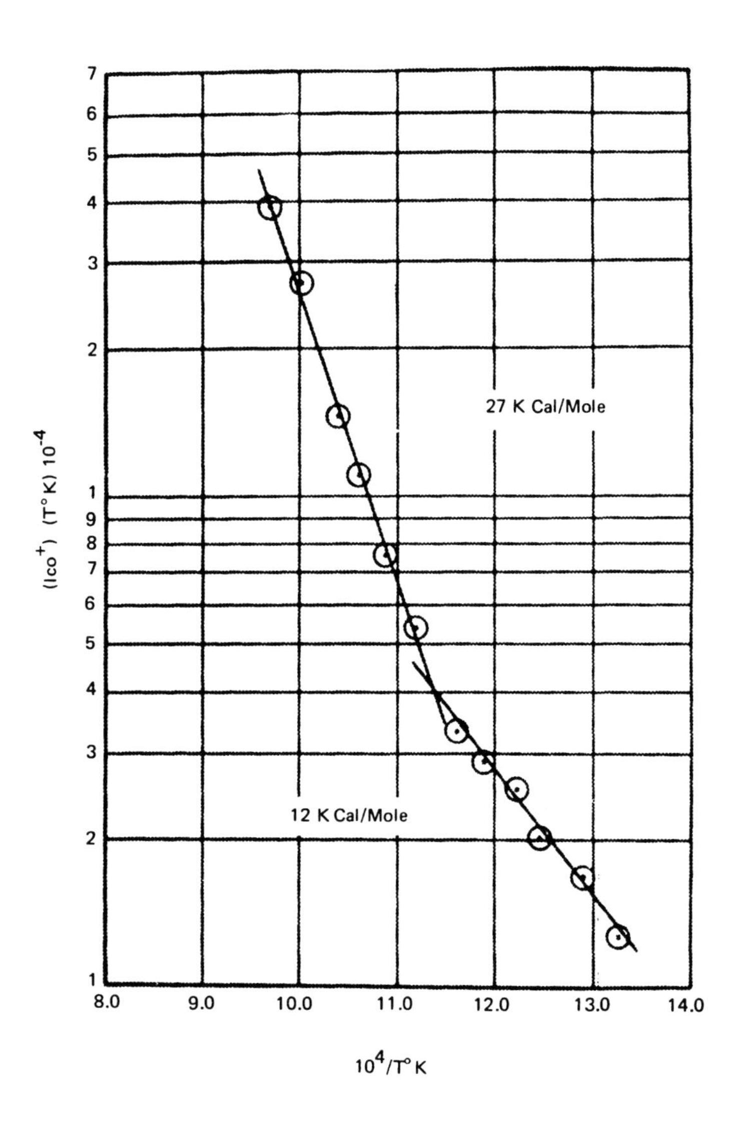

Figure 12. Energies of release for carbon monoxide from DTA data.

The results on the spark source are in agreement with the data presented by Morrison (Morrison et al. 1970). This work was done totally independently with another instrument.

WASSERBURG: The thermal contact in some samples could cause problems. What is effectively a pyrolysis type experiment may depend upon the thermal contact and if you embed this, part of the things get heated one way in one minute and another part gets heated another way.

ORÓ: This is done progressively. I do admit that you can probably have deviations.

Figure 11 shows the DTA curve, with two points, 1134°C and 1211°C. These are pretty close to the 1210°C reported for the separation of the ilmenite, and the 1140°C for the separation of pyroxene, troilite and metallic iron. I feel comfortable that these values are almost identical to the other values reported in the literature.

Figure 12 is a plot of the intensity for the carbon monoxide ion versus the reciprocal of the absolute temperature. From these measurements we have an idea of the energy involved in the release of the carbon monoxide. As shown here it varies from 12 to 27 kilocalories per mole. This is for two ranges of temperatures at which these measurements were made.

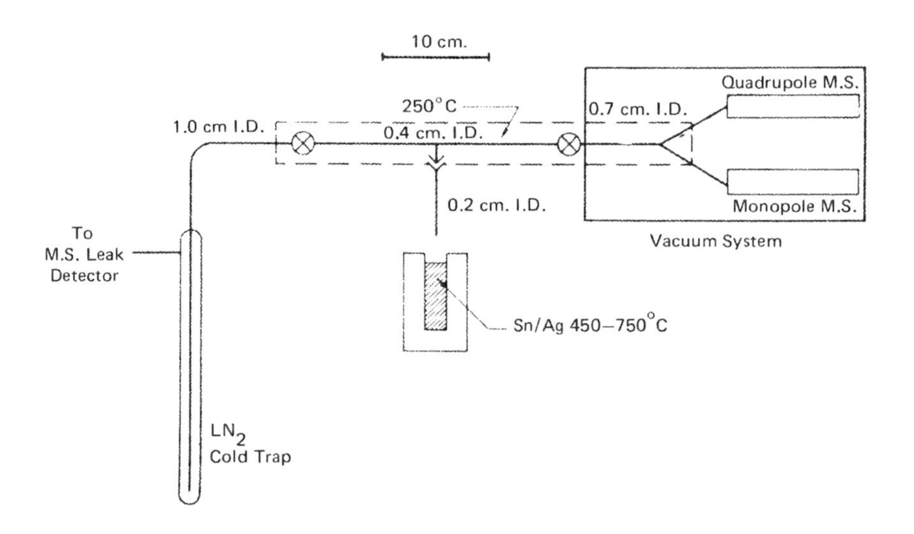

Fig. 13. Thermal volatilization quadrupole mass-spectrometer system.

These values are compatible with different forms of CO binding and release, including chemi-adsorption of CO, decomposition of carbonyl derivatives, or reaction of elemental carbon with iron oxide.

MILLER: Is that the sort of carbonyl you mean, rather than acid aldehyde, which is a carbonyl?

ORÓ: That is pure CO; I am not referring to any organic compound, but rather to the combination of CO with iron or other metals. These data together with those on the release of gases force me to conclude we probably have several sources of carbon monoxide, but right now the data does not warrant a complete discussion so I am leaving it entirely open as being a free gas, a derivative–metallic complex–or a reaction product.

Figure 13 shows a diagram of the quadrupole system. A lunar sample of about 200 mg is put in a small stainless steel tube–0.2 cm i.d.–which is heated by means of Sn/Ag alloy bath, and the mass spectra are measured.

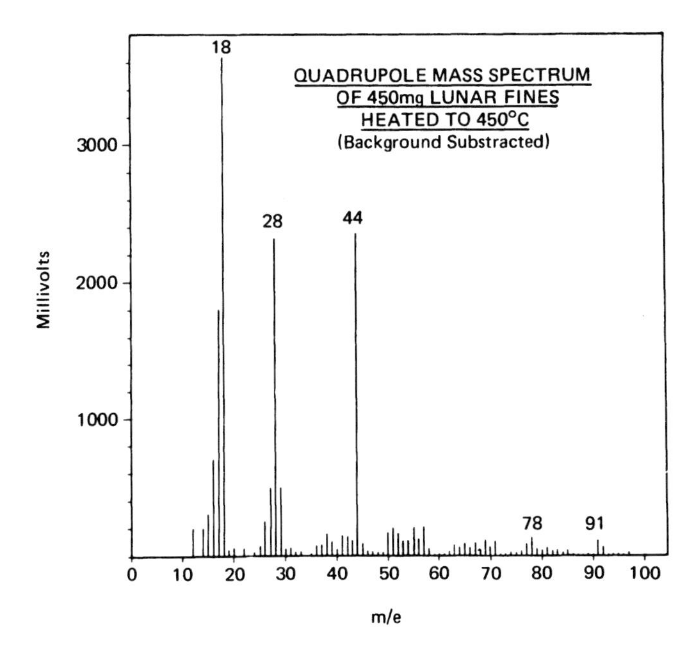

Figure 14. Quadrupole mass spectrum of lunar material volatilized at 450°C.

Here in Figure 14 you see the 44 mass for carbon dioxide, and the 28 mass for both the N_2 and the CO. There are several other peaks. For instance, with regard to mass 27, it would be of interest to know if it were definitely hydrogen cyanide. We cannot say now. There are traces of 78 and 91 peaks corresponding to benzene ion and the tropillium ion from toluene.

RICH: Do you think those larger ones are real?

ORÓ: We think carbon monoxide, carbon dioxide, and nitrogen are real, and that is where I will stop. I don't know what the others are. There also is a peak at mass 18 that corresponds to water. We think this is terrestrial water.

ORGEL: Can you tell us how big the spectrum of the blank was? Is it a difference between two quantities of the same order?

ORÓ: I think it is at least two orders of magnitude.

ORGEL: For benzene and for toluene?

ORÓ: I think it is essentially the same difference, two orders of magnitude. I do not recall the exact figures for benzene and toluene.

Figure 15 shows the system for combined gas chromatography-mass spectrometry. Here [pointing to the reactor] we put a pyrolyzer or a small vessel to carry out a thermal volatization for a chemical reaction. Figure 16 shows two types of these reactors. One is essentially a quartz tube with a heating block that can go up to 750°C. The other is essentially a glass vessel where the lunar sample and the reagent–sulfuric, phosphoric, or hydro-

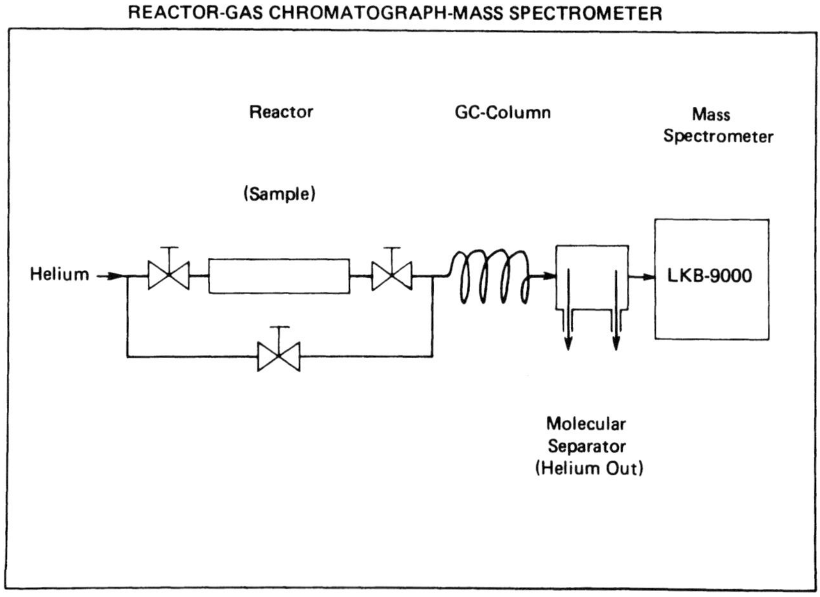

Figure 15. Diagram of the reactor-gas chromatographic-mass spectrometric system.

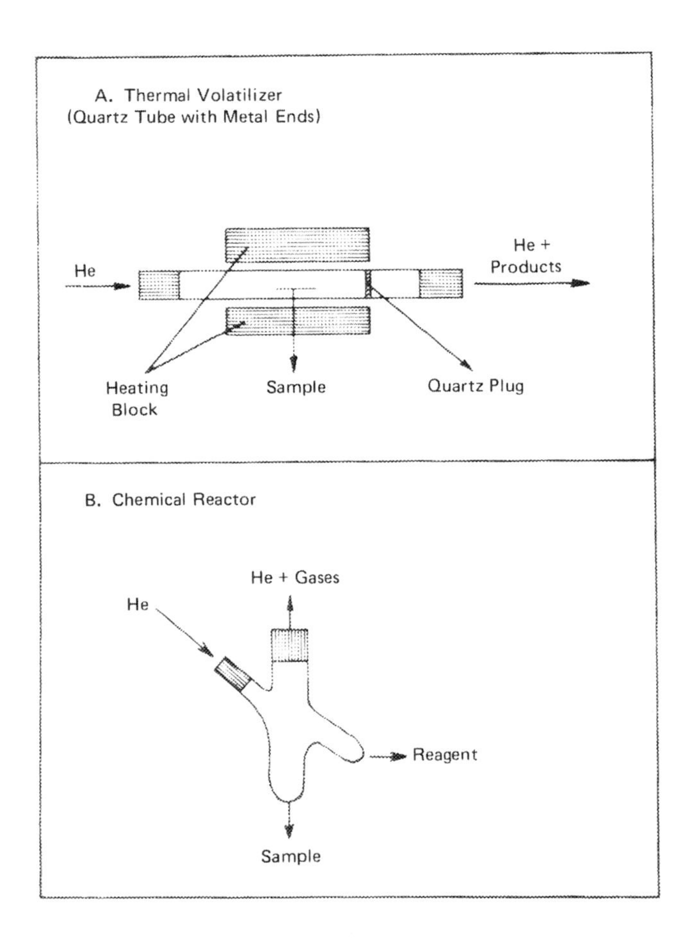

Figure 16. Reaction vessels used for thermal and chemical treatment of samples. A. Thermal volatilizer. B. Chemical reactor.

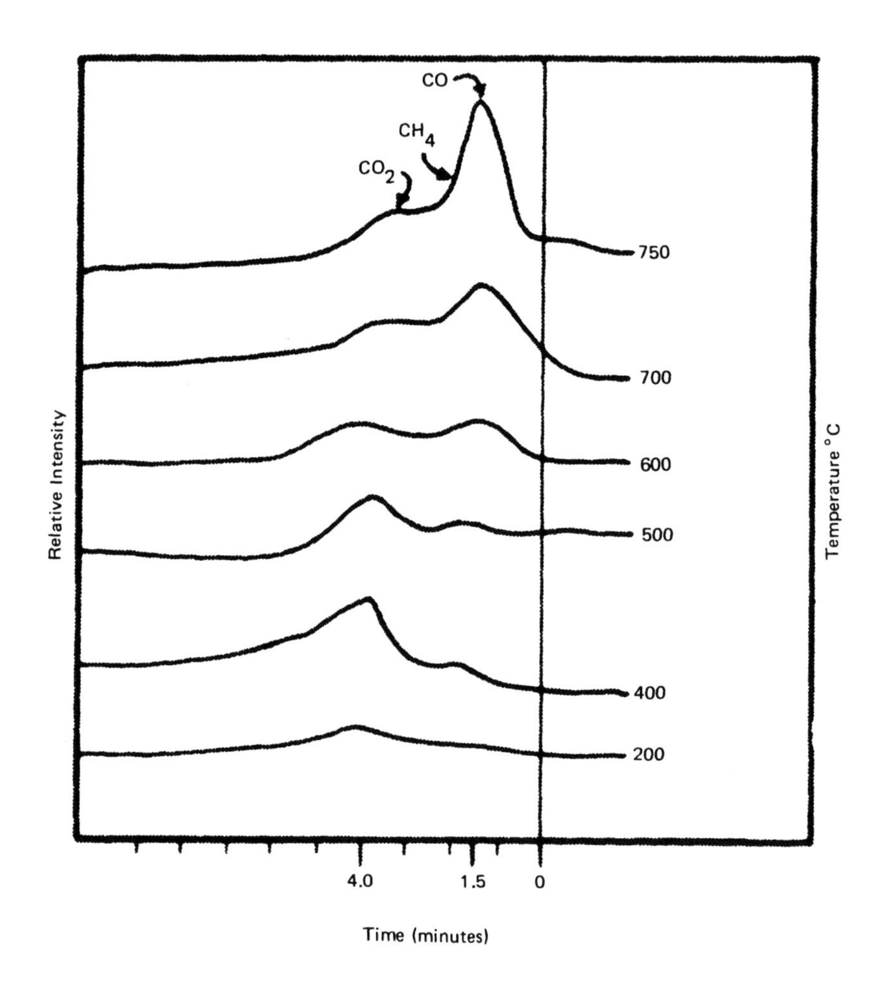

Figure 17. Gas chromatograms of CO, CO_2 and CH_4 obtained by sequential heating of a 1.5 g sample of fines. Semicontinuous mass scans of the gases were taken. CH_4 appears in the shoulder of the CO peak.

chloric acid–are placed in separated compartments and are eventually mixed and allowed to react. After a certain reaction time the volatile products are formed into the gas chromatograph.

Figure 17 shows several gas chromatograms obtained by stepwise heating of the lunar fines. Here [pointing to lower chromatogram] we heated the sample to 200°, allowed the material to go into the gas chromatograph and analyzed it. We find this very small peak corresponding to CO_2. Continuous scanning of the peak in the LKB instrument shows that this is a CO_2, and only CO_2. If we go from 200° to 400°C, the CO_2 peak increases. It has a maximum at about 400° and then another peak starts to appear which corresponds to CO. As heating is continued, the CO_2 peak really does not increase much more but now the CO peak takes over and increases almost linearly up to 750° –the maximum temperature obtained with this apparatus.

From 400°C up, if you scan the shoulder of the CO_2 peak you start to measure methane. These data, obtained by gas chromatography-mass spectrometry, of the products formed by thermal treatment of the sample confirm and extend the observations made by the high resolution mass spectrometry and low resolution quadrupole mass spectrometry. The amounts of methane are very small indeed, but they are there. This agrees with work of other investigators.

Figure 18 shows the results of three experiments carried out by adding acid to the fines. The gaseous products formed were separated by gas chromatography and analyzed mass spectrometrically. The major product is hydrogen sulfide in line A. We trapped the hydrogen sulfide by means of colloidal copper so we could see the peak of CO_2. Here is the CO_2 peak and essentially nothing else in line B. We trapped CO_2 by means of beads of sodium oxide and sodium hydroxide. After trapping or absorbing H_2S and CO_2, gas chromatography revealed the presence of a large peak of methane and smaller peaks of acetylene, ethylene, and ethane in line C. This is an indication that perhaps carbides–some produce methane rather than acetylene as a major product–are present. Other investigators have shown that cohenite and some iron nickel carbides are present in the lunar rocks.

MILLER: Can you tell us how much gas is coming off?

ORÓ: Only very roughly. These data are consistent with the values of 1,000–4,000 parts per million of sulfur that we have found by spark source mass spectrometry, and the amounts of CO_2 are a fraction of the total carbon. Very approximately we have on the order of 10 to 20 ppm of CO_2. I am not prepared to give the ppm of methane.

MILLER: About 2, 1?

ORÓ: Possibly not more than 5-10 ppm.

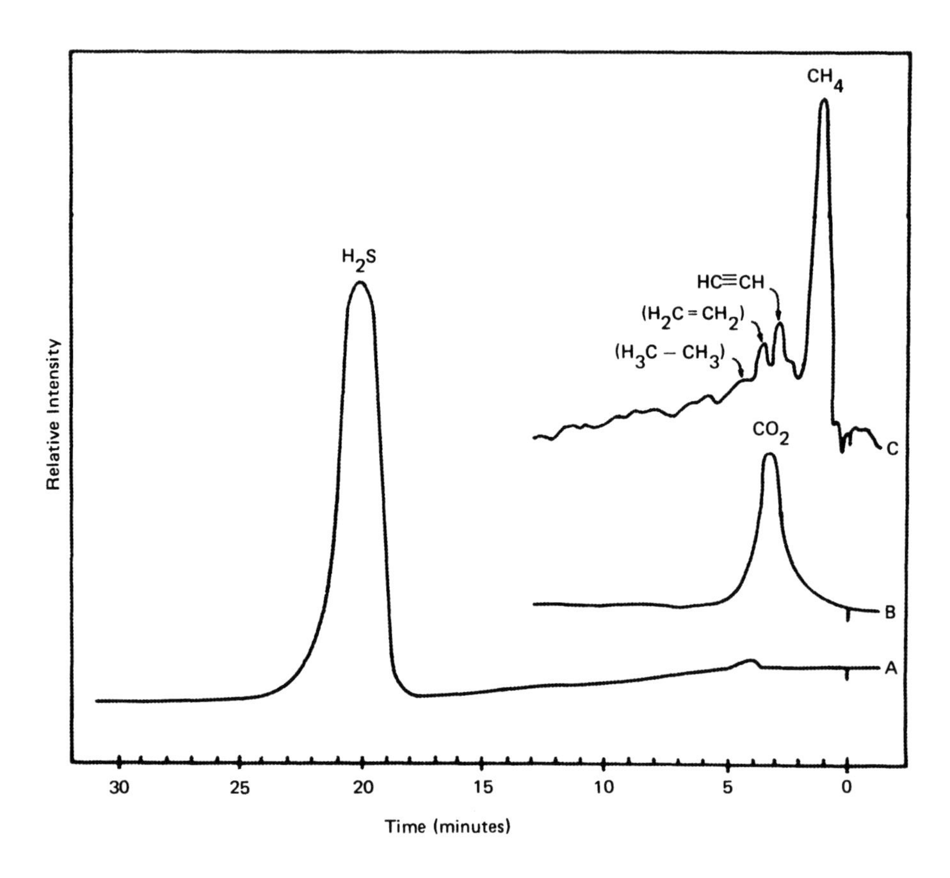

Figure 18. Gas chromatograms of compounds released by acid treatment. Semicontinuous mass scans of the gases were taken. A. Phosphoric acid. GC column (1.2 m long) of 200 mesh Porapak Q operated at 0.82 kg/cm² of helium. B. Sulfuric acid. GC column (0.6 m long) of 200 mesh Porapak Q operated at 0.35 kg/cm² of helium. C. Hydrochloric acid. Same conditions as in C.

WASSERBURG: In the acidification experiment it is conceivable that the CO_2 observed is CO_2 released from the phosphate.

ORÓ: CO_2 released from the phosphate as a phospho-carbonate?

ORGEL: Carbonated apatite.

WASSERBURG: Yes. There is about 0.2 of a percent of phosphorus and we know that the phosphates are roughly 90 percent stoichiometric without the CO_3 group in.

I did not know that any aragonite had been reported.

ORÓ: It has been indicated by Mossbauer spectroscopy (see Gay et al. 1970).

Figure 19 shows 0.5 ppm or less of total organic material obtained by heating up to 480°C. It is not known whether these traces of organic compounds belong to the lunar samples or are derived from terrestrial contamination.

Figure 20 shows something editors don't want because they say it shows essentially nothing. However I think it is very important if we have to establish a level at which we see essentially nothing. The tracing B [pointing to upper illustration] refers to the total amount of organic extractable compounds found in the bulk of sample 86,16, of which 104 grams were received in our laboratory. This was done twice with 1 gram each time by 2 independent investigators, one of whom has spent at least 3 or 4 years analyzing meteorites. Both of them obtained the same result–the total integration of the curve was equal or smaller than the integration of the blank that we had run. There is less than 0.001 ppm of organic extractable compounds. To me, this means there is none.

Tracing A is interesting. This corresponds to a small amount of lunar material taken from the top layer adjacent to the screw cap of the aluminum sample container as it was received in our laboratory. This cap has a liner of teflon. It shocked us to find this organic compound–not in very large amounts, 1 ppm–of molecular weight of 150. We do not know yet what it is. Possibly it has some sulfur in addition to carbon. It seems some-

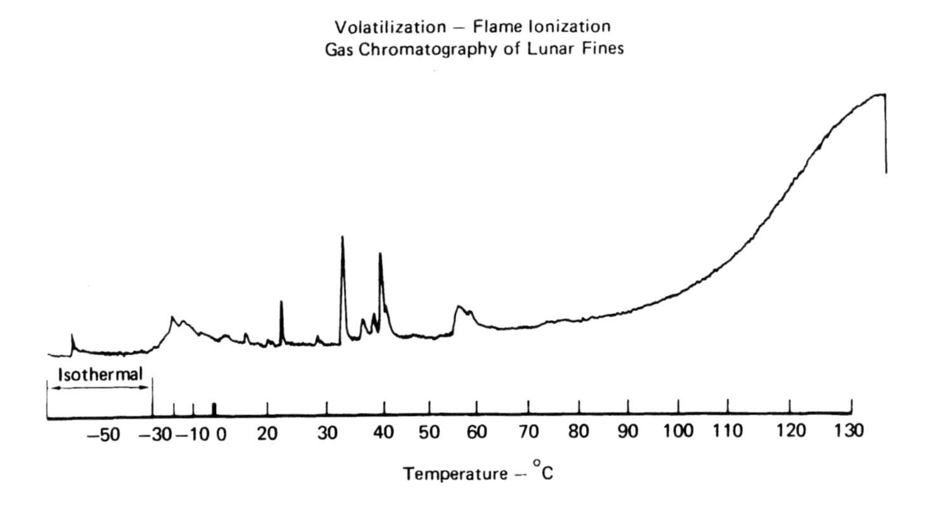

Figure 19. Flame ionization gas chromatogram obtained by pyrolysis of fines at 480°C, condensing the volatiles in a column at subambient temperature, and program-heating this column to 130°C.

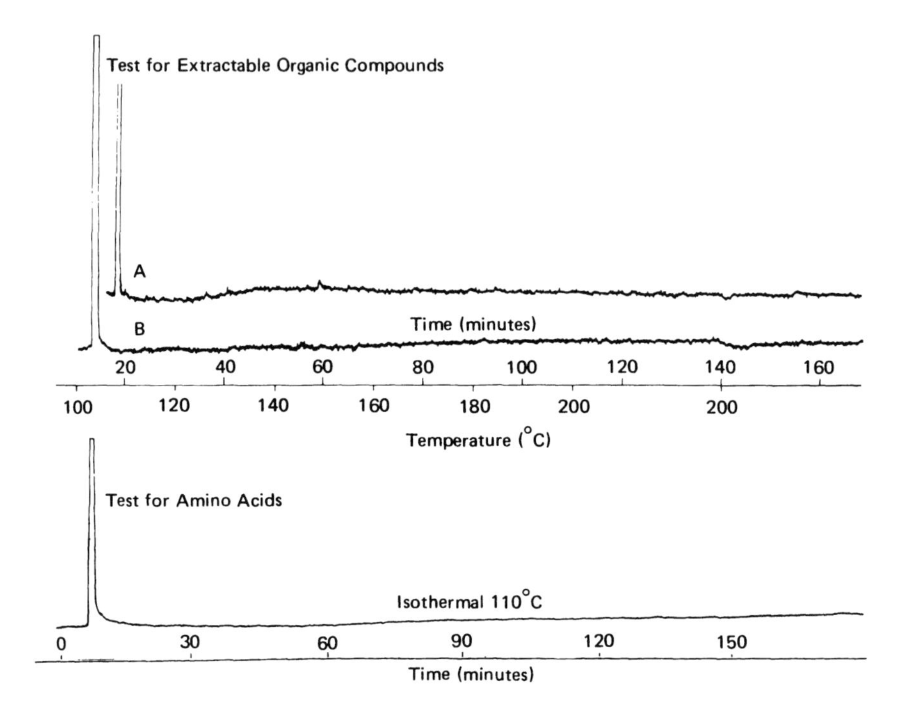

Figure 20. Gas chromatograms obtained for extractable organics and amino acids. A. Single component (X) extracted from 0.123 g of fines from the top layer of the sample container. B. Extraction of 1.105 g of fines from the bottom layer of the sample container. C. Amino acid N-TFA isopropyl ester derivatives of the hydrolysate from 1.4 g of lunar fines.

what similar to the compound we detected in extracts of the York mesh from Apollo 12. Whether the compound came from the York mesh of Apollo 11 or came from the teflon liner, we do not know. We consider it is a contaminant. [It has recently been identified as diisopropylsulfide.]

MARGULIS: What is York mesh?

ORÓ: York mesh is a spongy metallic net made probably of an aluminum alloy. It is used as packing material in the Apollo boxes to prevent the rocks from breaking to pieces.

The lower tracing [see C] shows there are no amino acids–at the 0.1 ppm level–and, therefore, there are no optically active amino acids either. If there had been optically active amino acids we would have been able to separate them as N-trifluoroacetyl isopropyl esters on an optically active phase–N-trifluoroacetylvalylvalylcyclohexyl ester.

We agree with Epstein and Kaplan that the $^{13}C/^{12}C$ isotopic values are positive–Table 4. Three runs were made following Craig's method and the values obtained varied from +14 to +18.5 per mil versus the PDB standard. We obtained no negative values. Perhaps this is because we were a little generous with the amount of sample used. However, we calculated in advance that we needed from 1 to 10 grams of sample and knew it would be silly to try to measure isotopic values using milligram amounts.

The three samples that are in the lower part of Table 4 gives values from +2.6 to +4.6 per mil. These values were obtained using a higher temperature method–1,400°C. We don't understand the variation. A simple interpretation is that the carbon–graphitic, carbidic, et cetera–released and combusted at higher temperature is less positive.

TABLE 4. TOTAL CARBON CONTENT AND STABLE ISOTOPE DISTRIBUTION IN APOLLO 11 BULK FINES SAMPLE.

Run No.	Sample Wt.	Combustion Method	Combustion Temp.	Total C (ppm)	^{13}C (per mil vs PDB)
1	11.0 g	Craig	800°C	40	+13.0
2a	10.9 g	Craig	900°C	110	+18.5
2b		Craig	900°C	11*	+14.1*
3	9.6 g	Fusion	>1400°C	60	+2.6
4	2.1 g	Fusion	>1400°C	210**	+2.6
5	2.9 g	Fusion	>1400°C	190**	+4.6

* *Additional CO_2 obtained from re-combustion of the same sample (10.9 g) after Run No. 2a.*

** *Samples 4 and 5 were fines passed through a 200 mesh sieve.*

We have indications consistent with mineralogical observations on the heterogeneity of carbon phases. This is the work I just referred to concerning the production of light hydrocarbons, presumably from carbides. Also, several micro-images for carbon and mass spectra of C_2 negative ions –m/e = 24–were obtained showing that the distribution of this element in the lunar fines and breccia is at times diffuse but can also be observed in more or less well differentiated patterns. However the significance of these data is not clear at present because of the newness of the method and the limited experience with comparable samples. The images obtained from the Pueblito de Allende meteorite showed several small–10 microns–spheroidal bodies of carbon. More work needs to be done with this technique on meteorites before it can fruitfully be applied to the study of the carbon phases in the lunar samples.

All the data presented here is summarized in Table 5. Two major conclusions may be made. First, organogenic elements are found in surface material retrieved from the Sea of Tranquility; secondly, the gases released with increasing temperatures are–in addition to hydrogen and the noble gases–

Table 5. Summary of data obtained by application of several techniques to the analysis of type D (fines) and type C (breccia) lunar samples from Apollo 11.

	Analysis	Description of Method	Measurements	Principal Analysts
1	Elemental	Spark ionization mass spectrography	41 elements including the organogenic H, C, N, S, and P	W.S.U., J.S.
2	EGA & DTA	Effluent gas analysis from 100-1400°C by high resolution mass spectrography and differential thermal analysis	Noble gases, CO, CO_2 and N_2 ($CO > CO_2$)	W.S.U., F.W., W.K.S. D.E.G.
3	Volatile and/or pyrolytic products	Stepwise heating to 750°C and quadrupole mass spectrometry	H_2O, CO (and N_2), CO_2, C_6H_6, C_7H_7 and traces of other organics	J.McR., W.S.U.
4	Volatile and/or prolytic products	Stepwise heating to 750°C and gas chromatography (Porapak column)-mass spectrometry	CO/CO_2 release vs. temperature CH_4 above 400°C ($CO>CO_2$)	J.G.
5	Volatile products from acidolysis	Acid treatment and gas chromatography (Porapak column)-mass spectrometry	H_2S, CO_2, CH_4 and other light hydrocarbons	J.G.
6	Trace volatile and/or pyrolytic organic products	Stepwise heating to 430°C and flame ionization gas chromatography (SE-30 column)	<0.5 ppm	R.L.L., C.J.W.
7	Trace organic solvent-extractable organic compounds	Flame ionization-high resolution-gas chromatography (Polysev column)	<0.001 ppm	J.G., D.W.N.
8	Amino acids	Flame ionization-gas chromatography	<0.1 ppm	E.G., J.I.
9	Carbon-13	Carbon isotopic ratio mass spectrometry	$\delta^{13}C = +2.6$ to $+18.5$ PDB	D.A.F., P.P., A.N.F.
10	Solid carbon compounds	Ion microanalysis	Carbon image from C_2^- ions	W.S.U., R.L.

mixtures of carbon monoxide, carbon dioxide, nitrogen and traces of methane. In the presence of water and on a planetary body of sufficient gravitation, these gases would constitute an atmosphere which, by means of solar radiation, electrical discharges, or high temperatures could give rise to the formation of precursors of organic and biochemical molecules. The results obtained from the moon are informative in this respect. In a sense it is preferable to have found these organogenic gases than to have found organic material *per se;* for they provide the step prior to the synthesis of organic molecules. What we are planning to do now is to submit this mixture of gases to a number of different sources of energy, with the hope that we can synthesize biochemical molecules.

STROMINGER: Juan, suppose you took some terrestrial rock from some place where there was minimum life; from the top of the Himalaya Mountains, or from the Sahara Desert, or the Antarctic, or any place on earth where the amount of life is minimal. What amounts of compounds of the type you have found in lunar rock would you expect in that type of terrestrial rock?

ORÓ: That's a very good question. I can answer it in two ways. Bill Rubey made precisely this point in his presidential address to the Geological Society. Gases obtained from the earth by either volcanic or magnetic processes are somewhat similar to the ones found here: CO, CO_2, N_2, methane, probably also H_2S, water, hydrogen.

What happens with a magmatic or other igneous rock that has a minimum of life? I don't know, but this type of work needs to be done. Unfortunately, we are studying the moon before doing this work on terrestrial rocks. If, as Rubey says, the earth produces these gases, then residual amounts of them will still be present in some of these rocks. I don't know how much, but we are actually looking into this particular question now.

HOROWITZ: Volcanic gases that Rubey mentions aren't necessarily juvenile; volcanic CO_2 could be of secondary origin.

ORÓ: This is the question we were just discussing. Wasserburg talked about a high reservoir. There is a high reservoir in the carbonatites which are considered to be igneous or primeval, or whatever word you want to use.

HOROWITZ: If you melt a real old Precambrian rock from the mantle– I have forgotten the geological name–

ORÓ: What happens? I am sure that if I take a carbonatite I will get CO_2 out.

WASSERBURG: Calcium carbonate–I should hope to hell you would.

HOROWITZ: If you melt igneous rock don't you get methane, reduced gases?

ORÓ: Such statements are in the literature and I am prepared to accept that some of these compounds are there in small amounts.

WASSERBURG: Although this isn't my field I have actually done these experiments for exercise and never published the results. The water-gas reaction dominates everything, and it is a most serious matter. Anytime water is involved and meteorites are analyzed, rather obvious effects of the water-gas reaction are seen. Basically, this is what you are usually studying including our great discovery of free hydrogen—which we had the intelligence not to publish because it wasn't as free as we would have liked.

MILLER: What, exactly, do you mean by "water-gas reaction?"

HOROWITZ: Carbon plus water.

MILLER: I don't think C plus water is water-gas. I think it is CO plus water or plus hydrogen.

Your first point on this business of heating things up is that much volcanic gas is secondary, and your second point is essentially what Wasserburg said: A very reduced set of gases like methane and water become unstable with respect to CO and CO_2 at high temperatures and are broken down–the methane to CO and CO_2 just by the shift of equilibrium, which is water-gas.

STROMINGER: What do you get if you compare the percent of terrestrial carbon on the earth with all its mass, including the atmosphere, with the percent of carbon on the moon?

ORÓ: I have not made that calculation but obviously the amount of lunar carbon is very small; I concur with Gene Shoemaker's statement that practically all the rocks have been degassed.

Of course all the sedimentary carbonate plus all the igneous carbonate in the form of carbonatites, plus all the carbon in the form of living organisms and fossils, added together and related to the mass of the earth—I really don't know what you'd get, maybe Gerry does.

WASSERBURG: There is no carbon in the atmosphere or in the biosphere. It is tied up in the oceans, the igneous and carbonate rocks.

ORGEL: Are there no carbides in the interior?

WASSERBURG: We don't have the foggiest idea.

ORGEL: So you can't rule it out?

WASSERBURG: The earth's surface has large reservoirs so the carbonates and the bicarbonate-carbonate stuff in the oceans can be assessed but with respect to the moon there are many calculations.

HOROWITZ: All the terrestrial limestones put into the atmosphere would yield a lot: about 20 atmospheres of CO_2 or 20,000 grams CO_2 per square centimeter. From what I have heard today, nothing like that is on the moon. The earth has been outgassed and the carbon has come from inside; the moon is not differentiated that way. I suppose if the moon were outgassed the way the earth is they might be comparable.

WASSERBURG: The calculation can't be made.

ORÓ: We cannot make a strict comparison of the percentages of carbon in the moon versus the earth. Even if initially the two bodies had the same carbon composition the carbon probably would not have been retained on the moon if it had been heated to, let's say, 1200°C. Because of the high gravitational field of the earth the carbon may have been retained in our planet.

MARGULIS: Given your best model for the moon, where do you think the carbon is?

WASSERBURG: Inside, in the rocks. We don't have a representative sample of the moon; we have some tidbits. From the fact that our rocks are anhydrous, can we say there is no water on the moon?

ORÓ: I want to mention an observation consistent with one of Carleton Moore (Moore et al. 1970). Apparently the finer the particulate material is, the higher the amount of carbon is. This may imply carbon of solar origin but other interpretations can be advanced. I think Carleton Moore measured up to 500 ppm for very fine particulate material. As I showed in the table we found 200 ppm for the material that was sieved and about 120 ppm for that not sieved.

WASSERBURG: What size was the sieving?

ORÓ: 100 mesh.

WASSERBURG: Most of the coarse material comes from crystalline rocks, so all you are doing is separating out stuff that you know doesn't have the carbon in it anyway.

ORÓ: That's right.

HOROWITZ: Let's call on Fred now. He has to leave very soon.

ORIGIN OF THE MOON BY CAPTURE

SINGER: I was asked to give a little theoretical perspective on the problem of the origin of the moon which I look at from the particular point of view of celestial mechanics. Although my viewpoint is different from most of yours, it is obvious that we need to discuss the problem from several viewpoints, for example: the geological, the celestialogical point of view, and the point of view of the moon rocks.

I want to remind you that the lunar orbit is influenced by the fact that the moon produces tides on the earth and these tides today influence the orbit of the moon to a certain extent. The moon is changing, expanding its orbit at the rate of about 1 inch per year, because of the tides which the moon produces on the earth. These terrestrial tides perturb the orbit of the moon.

A number of people including Darwin have extrapolated backwards in time to determine the moon's orbit in the past. Everyone agrees with these calculations up to a certain point. Agreement exists down to a distance of approximately 5 earth radii. The moon is now a distance of 60 earth radii and most agree with the results of these calculations down to about 5 earth radii and then differences—which I will summarize very briefly—develop.

RICH: When was the moon at 5 earth radii?

SINGER: I cannot do justice to this in a short discussion. The time scale essentially for the total orbital development of the moon is usually based on the rate of tidal dissipation observed today, which provides a single constant. This turns out to be about 1.7 billion years.

I will very briefly describe three major theories of the origin of the moon and essentially eliminate two of them. I warn you that each theory still has strong partisan support, but the partisans for the other theories are not here tonight. It would be unfair of me to subject you to my point of view without cautioning you that there are other views that are valid.

The first of the three theories for the origin of the moon is fission; this has been advanced by several people including O'Keefe (1970) in its most recent form. This is the process whereby the moon is separated from the earth, presumably sometime after the original earth was formed. This separation of the moon from the earth takes place after the core of the earth has formed. The formation of primitive earth is followed by the formation of a core in the earth which produces an instability of rotation. This results in the separation of a chunk of matter which becomes the moon.

The second idea of the origin of the moon is the accumulation or coalescence of small chunks of matter which happened to be in orbit around the earth or were left in orbit after the earth was formed. This coalescence theory has been proposed primarily by Opik (1962), and more recently, for separate reasons, by Ringwood (1970) and by Cameron (1970).

The third major idea of the origin of the moon is the so-called capture hypothesis which postulates that the moon was formed somewhere else in the solar system as a complete body, and then later was captured by the earth to eventually end up in the present orbit. I believe this was first proposed in sensible form by Gerstenkorn (1955) and then was taken up by Alfven (1965) who added to Gerstenkorn's suggestion.

All of these theories have been rejected at some time or other for good reasons. Goldreich (1966) presented sound objections to the fission theory that were based on celestial mechanics arguments. Major objection to the coalescence hypothesis also has been given by Goldreich, on similar celestial mechanics arguments.

The major objection to Gerstenkorn's capture theory had been given by MacDonald (1964) who had shown that the earth would be excessively heated by a lunar capture. So we are left with all three major theories rejected.

My point of attack has been on capture theory. I examined MacDonald's recalculation of Gerstenkorn's original work and found some things lacking. I came up with a different result (Singer 1968) which removed the objections MacDonald had raised to the capture theory in the first place.

MacDonald's work–probably the best thing to read on the subject of capture–suggests the moon would have been captured by the earth from a so-called retrograde orbit. This means that the lunar orbit would possess angular momentum opposite to that of the earth. Since now the moon has orbital angular momentum which is in the same direction as that of the earth, it means the moon flipped around. This cataclysmic process would result in such tremendous changes in the earth's rotation and in the energy that is dissipated in the interior of the earth that the earth would vaporize as it captured the moon. Of course this has not happened. So, on this basis MacDonald rejected the capture theory.

But doing this calculation with a physically more realistic theory of the tides–there is a fundamental difference in the tidal period I use–the moon, as we go back in time, ends up with a direct prograde orbit and, therefore, the changes in the angle of momentum of the moon are more reasonable and changes in the rotation of the earth much less. Therefore, the energy dissipation of the earth's interior would be greatly reduced.

This seems to provide a way of getting around certain physical objections to the capture theory. I have to admit capture is still very improbable, but the moon is there and its present position must be accounted for.

What are the physical consequences if the moon was captured by tidal friction, as I propose here? There would be consequences to both the earth and moon, and geology comes in because we can see whether or not the consequences have occurred. In the first place, I can reject this time scale completely by assuming that the coefficient which was used in the fictional type of theory—which applies to the situation to date—has not been constant. It is artificial to assume that tidal dissipation has remained constant throughout geological times.

RICH: How would tidal friction be affected by continental drift?

SINGER: It is. It is affected by the condition of the continents, by the way the shoreline runs, by many things. Celestial mechanics tells us nothing about the time scale, but we can be fairly sure that it is not 1.7 billion years. There is nothing in the earth's history that shows any correlation with 1.7 billion years.

WASSERBURG: When it was 0.9, it wasn't any good at 0.9 and now that it is 1.7 it is no good at 1.7. Since it goes like 1 over r to the 6th or the 9th, or something, the time scale is fixed by the long distance interaction and when it gets in close it goes like gangbusters. That is a lousy way to date the earth and nothing has happened to us, so this idea is out.

SODERBLOM: What kind of correlation are you looking for?

SINGER: If the moon had been close to the earth 1.7 billion years ago, something in the geologic record, some cataclysmic effects in the earth should have been seen.

SODERBLOM: Such as anorthosites in the Precambrian record?

WASSERBURG: You are talking about a great kooky thing. The moon was very close to the earth. Where it came from God only knows at this magic time.

SINGER: Celestial mechanics tells nothing about the time scale. Time scales must come from geology; they cannot be manufactured from celestial mechanics. I mention this because people have taken this time scale very seriously as a basis for criticism of the celestial mechanics work.

LEOVY: Is the actual process that one is lumping into "viscosity" or whatever so poorly understood theoretically that—

SINGER: It is understood well enough to know it could not have been constant throughout geology.

LEOVY: Do you know how it would have had to vary, or is this ad hoc?

SINGER: No.

ORGEL: Can you tell whether the viscosity was bigger or smaller?

SINGER: No. I suspect smaller. Again on the continental drift, if there were two major continental masses and if there were oceans–

WASSERBURG: I don't understand. Surely if the calculations have any sense, the time scale, regardless of the means of origin of the moon, must be 4.6×10^9. That must be the solution to the problem.

SINGER: That is my solution, but it doesn't come from celestial mechanics.

WASSERBURG: You impose this time scale and change the viscosity to get it. I don't care whether the moon originated by fission or by capture or by condensation, you must ask what happened prior to the time when you go back to equal zero. The only way to avoid this is if you bring the whole thing back and tie the knot at the beginning so that you avoid an intermediate catastrophe.

ORGEL: Does the viscosity depend only on the distribution of land masses or does it also depend on their movements?

SINGER: The dissipation depends very much on where the moon is, but this is one thing we can calculate. There is a disposable constant which depends on such things as the disposition of land masses, the way the oceans interact with the shoreline, and so on. These are not known, but we can extrapolate into the past.

ORGEL: Do you put your understanding of where the moon is into the calculation?

SINGER: Celestial mechanics show you the shape of the lunar orbit at successive time intervals in the past without giving you the time scale. I play the time scale like an accordion. It can either be stretched or compressed, but its value is not available from celestial mechanics.

RICH: T equals zero must be 4.6 billion years ago for the first two theories, but need not be true for capture, provided it can be done in a non-cataclysmic fashion.

SINGER: I don't see how the moon could be captured in a non-cataclysmic fashion, but there is an idea that Urey and MacDonald (1970) proposed. Let me describe it quickly.

The earth is formed here and the moon is formed somewhere else. The earth is surrounded by a bunch of bodies at a distance of about 40 earth radii. Now, the moon happens to come by, sweeps up these bodies, and ends up in a circular orbit.

WASSERBURG: An awful long time ago.

SINGER: Yes.

RICH: There were lots of little moons in the beginning?

SINGER: Yes.

MARGULIS: And no cataclysmic effect at all?

SINGER: Not on the earth, because the moon is at too great a distance to influence the earth. I see no physical objections to this idea, but it involves a lot of assumptions.

Anyway, if one extrapolates and uses this new tidal theory in a consistent way, ultimately the effects of solar force are included and the higher order terms in the tidal theory are allowed for. There are other things to be done—I did these calculations last fall in Houston. I will give you the picture I ended up with right from the start.

The earth is in its orbit around the sun. The moon would have to have been formed in an orbit very similar to that of the earth—roughly the same distance from the sun, roughly the same eccentricity and the same inclination. If the moon catches up with the earth or the earth catches up with the moon it doesn't matter, every few years they come close together, and the most likely thing to happen in this kind of encounter is exactly nothing. The moon may describe a little orbit around the earth and then escape again, but that is the beauty of the thing: Nothing happens, so every few years the moon can try again.

Next most likely and far less likely than nothing to happen is for the moon to hit the earth and disappear. The least likely thing to happen is for the moon to approach the earth with the right impact parameter in just the right way so that the moon will spend a great deal of time near the earth and would eventually escape, according to the so-called restricted 3-body problem, except for the tidal interaction with the earth which removes energy from the moon. It then becomes permanently captured.

In order for this to be plausible, this whole process had to occur very shortly after the earth and the moon were formed, something like 4.5 billion years ago, not much later.

ORGEL: Why?

SINGER: The moon can't be stored anywhere. If the moon is in an orbit quite different from the earth the probability of capture is zilch.

RICH: You mean that it would be unstable as far as collision?

SINGER: No.

WASSERBURG: It is a cold storage problem: Where the hell do you keep it?

SINGER: How do you keep it away from the earth?

RICH: That is what I mean—unstable in terms of collision?

WASSERBURG: Suppose you wanted to capture the moon *yester-*

day; where would it have been stored between 4.6 billion years ago and yesterday?

RICH: The hypothesis is that it is in the earth's orbit to begin with.

WASSERBURG: That is an assumption.

SINGER: To make the capture a little more probable, the moon is put in an orbit like an asteroid which intersects the earth's orbit and is never captured.

RICH: Let's assume the moon is in the earth's orbit: How long will it remain in that orbit until capture?

SINGER: A very short time; let's say Y billions of years.

RICH: Why?

SINGER: Because every 2 years it encounters the earth. Let's assume the probability of capture is 10^{-6}, and the moon encounters the earth every 10 years; then it would be captured in 10 billion years.

ORGEL: Is it not the continuous deformations of the orbit that makes capture progressively less likely, so that it becomes virtually impossible? Couldn't you choose a position to make the expectation time of capture as long as you like?

SINGER: But it does get captured.

ORGEL: Are you suggesting a small family of orbits which inevitably leads to capture, and another family of orbits which inevitably leads to no capture, and no intermediate orbits which give lower probability of capture? Why doesn't it go continuously from one to the other so there is always some orbit which will give you a probability of capture which would take any arbitrary time?

SINGER: You can, but the probability becomes so small that one doesn't worry about it.

RICH: But how do you explain the existence of both earth and moon in the same orbit?

SINGER: That is a separate problem. You have to assume that the earth and the moon completely formed along with other planets. If you want to be more speculative–assume there were several moons, all of which have disappeared on impact on the earth, simply because impact is more probable than capture.

MILLER: In a typical orbit, what is the probability of capture versus collision with the earth?

SINGER: This I have only estimated to be around 1 to 10^4.

MILLER: Aren't you in favor of collision?

SINGER: Yes, collision is much more probable.

RICH: Aren't you getting rather close to the Urey-MacDonald picture now? You have several moons in the same orbit as the earth and eventually they get trapped in earth orbit.

SINGER: No, they don't get trapped in earth orbit.

RICH: Eventually they collide. Aren't all these several moons coalescing?

SINGER: No, no.

RICH: You are just capturing one?

SINGER: I am multiplying the improbability.

WASSERBURG: The best hypothesis was one that Tommy Gold–from Cornell University–put forth once at a conference in Jasper's Institute.* He got up and in one of his less wild moods began to rant and rave and said "Consider a random moon in a random sky." That is as close to an accurate statement of the problem as one can get. As Fred pointed out, all the calculations depend upon the details of some dissipation mechanism at the end. Capture is a very complicated problem.

SINGER: Anyway, this hypothesis has to be explored by looking at the consequences of this capture on the earth and on the moon, and this is what we try to do. From the celestial mechanics point of view, the virtue of capture is that it requires no extraneous assumptions. If this calculation is made correctly–and this must be judged by looking at how it was done–then the moon at one time must have had a highly centric orbit, nearly parabolic orbit, indicating that it started with this kind of orbit around the earth, which indicates capture.

Celestial mechanics, unfortunately, only defines an evolutionary path along which the moon could have been inserted anywhere. Let us assume the moon actually went through the complete evolutionary path and was captured. The time of this capture was very early in the history of the earth, about 4.5 billion years ago. The effects of this capture turn out to be most interesting and quite clearcut. The earth would have been heated by tidal friction which amounts to an average of about 1.5×10^{10} ergs/gram. We can be precise about this because we can calculate the change in the angular momentum of the moon during capture, which is obviously given. The earth must experience a corresponding change in angular momentum. To conserve angular momentum, the earth must despin and the energy that comes from the kinetic energy of rotation of the earth must be dissipated in the earth's interior. It is a quite straight-forward calculation. This number is very interesting because it is exactly what is required to heat a gram of silicate material to about 1,200° and melt it. The inevitable consequence of lunar cap-

*Institute for Space Research, New York City.

ture is to induce some melting–not necessarily uniform, of course–but some melting of the earth early in the earth's history.

MILLER: Can that 10^{10} ergs per gram possibly be deposited in the surface water?

SINGER: No, not if we assume there were no surface water.

HOROWITZ: This is what brings about the differentiation of the earth?

SINGER: Assuming what I call the classical model of the earth–let's say the Rubey model–where the earth was essentially bare of any atmosphere and water to begin with–the present atmosphere and oceans, of course, are of secondary origin.

MILLER: If there were water, does the energy get deposited in the interior of the earth or in the water?

SINGER: Much energy would have been deposited on the surface and there would have been less addition and less heating in the interior of earth, right.

The effect of this would be to include melting; the iron would then try to go to the center of the earth to form a core. This gives you an explosive, progressive melting effect in the earth resulting in a very rapid release of energy in the earth's interior and a very rapid perturbation of the core. Volcanism on the earth would start very shortly after the formation of the earth because of the capture of the moon. This is the major hypothesis that I put forward here.

The consequences are the formation of atmosphere and oceans on the earth 4 billion years ago, maybe earlier, and from the point of view of the origin of life the consequences are interesting: Prebiotic materials could have formed at least 4 billion years ago.

RICH: They could have been formed earlier, too, but just pyrolyzed afterwards.

WASSERBURG: Sure–the problem is now solved twice, since we don't know how to do it once.

SINGER: It is generally assumed by geologists–or it was assumed by geologists up until Wasserburg–that the earth's-mantle material was chondritic and contained certain amounts of uranium, thorium, and potassium. Those values of radioactivity had always been used to calculate how long it would take for the earth to melt and form a core. We are not sure about this anymore. If you want to say something about this, Gerry, please be my guest, but many now feel it might take as long as 2 billion years to melt the inner material and form a core. This would clearly be inconsistent with the geological and biological data we have concerning the existence of atmosphere and ocean on the early earth.

SODERBLOM: The melting of the earth and the differentiation thereof –is that not a function of the accretion rate of mass into the earth?

SINGER: It could be a function of the accretion rate if the accretion rate is rapid enough. It has probably not been rapid enough because the atmosphere and ocean are of secondary origin.

WASSERBURG: But there are several reservoirs of thermal energy; a nice part of the capture hypothesis is that another energy source can be juggled with, and therefore it is quite appealing, but I think the point is well taken. With just reasonable accretion rates, you can get a good bit up to heat the temperature.

SINGER: No. Not unless extreme assumptions on the accretion rate are made.

HOROWITZ: Venus has a dense atmosphere and it has no moon, so musn't there be another mechanism?

SINGER: We will come to that later. Mars does not.

SODERBLOM: You were telling us what the extreme assumptions on the accretion rate were that precluded it.

SINGER: The problem with accretion is that when you have lots of small bodies to begin with and rapid accretion rates, the gravitation potential of the earth is too small to give energy dissipation on impact. By the time the earth has grown, its gravitational potential is large and the accretion rate is very small because by then most of the bodies that formed the earth have been swept up. During accretion the energy is released on the surface of the earth and, therefore, would tend to be radiated away except for that amount of energy trapped so it depends very crucially on the rate. If the rate of accretion is not very rapid but instead reasonably slow, the energy is radiated away into space.

WASSERBURG: An infinitely small rate of accretion means a cold planet. Intermediate rates of accretion, which are not enough to volatize the whole into an ion gas–in which you can get about 600° Kelvin per gram–should be relatively easy to achieve without an anomalous rapid accretion plus the thermal heat which might provide about the other half. The whole thing has been robbing Peter to pay Paul all along. If you want to make the earth cold, you will have a hell of a time.

SINGER: Although it is not easy to make the earth cold, it is also very difficult to make the earth just 600° per gram. The accretion rates must be very delicately adjusted. Because it implies the molecules escaped at just a very definite rate, I don't think the evidence supports this. I don't think this is possible.

The major crucial point of interest in this particular theory of the origin of the moon is that the moon would be heated at the same time as the earth. Actually because the same mechanism acts to heat it, the mechanism is a little different and, to be more precise, in the case of the earth the heating takes place because of the dissipation of rotational kinetic energy in the interior that leads to the rapid formation of the core and then the onset of volcanism. This evaporates and exhausts gases on the earth's surface and results in the formation of atmosphere and oceans. It is not the rotation of the earth that causes the lunar heating but a radial deformation, a sort of pumping as the moon initially has an eccentric orbit and then acquires a circular orbit.

Because it depends on the elastic parameters of the moon I can't calculate this heating exactly, but reasonable assumptions can be made. If these elastic parameters are similar to those for the earth, then the prediction is that enough melting of the moon to cause some large-scale differentiations occurred. There must have been considerable exhaustion of gases at that time, too. This primitive lunar atmosphere and hydrosphere would have evaporated into space because of the small gravitational field.

For you, interested in the origin of life, the interesting question is whether or not the atmosphere and hydrosphere on the moon survived long enough to form biological material. Apparently the answer is no, judging from the results so far. Since the observations are on lava, which has been released from the interior of the moon more recently, one must be cautious in his interpretation. The picture I have of lunar history may be a little different, as follows. The moon was captured about 4.5 billion years ago. It was heated, large-scale differentiation took place, a general crust and an interior somewhat different from the crust were formed. The crust was distorted with respect to the center of the moon, probably by the closeness of the earth and its gravitational field. The maria were formed at that time or shortly thereafter, but lava flows took place at various times during the history. The Apollo 11 lava flows apparently took place 3.6 billion years ago; the Apollo 12 lava flows may have taken place some other time. Each mare may have its own dating for lava flow. The maria basins may be of the same age, much older than the lava flows which have taken place subsequently.

The lava of course was very hot, coming into contact with the vacuum it was completely degassed and it contains no water anymore and very little of the other gases. This is what one seems to find.

MARGULIS: Do these lava flows occur intermittently during the entire history or just during the first billion years?

SINGER: I don't know. As far as we can judge, during the first billion years and maybe more recently. It is possible that there is still a reservoir of younger material.

MARGULIS: Once the moon gets into the regular orbit are there any more changes in the earth-moon system? After the first billion years, does anything happen of interest that might be seen geologically?

SINGER: No, because the capture of the moon and the initial-orbit changes take place very rapidly. In the matter of a few thousand years the moon is brought in from its initially hyperbolic orbit into a circular orbit which starts to expand. Most of the 4 billion years or more is spent for the moon to go from a distance of about 5 to about 60 earth radii.

SODERBLOM: How long does it take to go from about 5 to 10 earth radii particularly in the presence of heating and the dropping very rapidly?

SINGER: This is being calculated; I guess now approximately 10^8 years.

SODERBLOM: You mean the moon can get out of any dynamic interplay with the earth in the order of a billion years?

SINGER: Yes, much less than that.

SODERBLOM: But if lunar-surface maria were found to be a billion or more years younger than those of Mare Tranquillitatis, how would that fit with your hypothesis?

SINGER: I differentiate between a mare basin and lava in a mare.

SODERBLOM: Aren't you talking about the fillings of the maria?

SINGER: Whenever there is a puncture or eruption of the floor, lava may be released. That is how the gas columns can be explained–they are caused by repeated releases of lava which become dense–when they become degassed they cause gravitational anomaly which we now call a masscon.

LEOVY: Is this 1.5×10^{10} ergs per gram dissipated in the order of a few thousand years?

SINGER: Yes.

LEOVY: Might you take a minute to reply to Norm Horowitz's question about the Venus atmosphere (p. 117)?

SINGER: What I have to say is very speculative and fairly ad hoc. Venus must have had some moons that have obviously crashed into the planet in such a way as to change its spin velocity considerably. This can be done by a kind of yo-yo effect leading to a great dissipation of energy in the Venus interior coupled with a despinning. Much gas must have been released. For some reason Mars has not had this problem.

ORÓ: A major problem you have is how to explain the different composition of the moon if it was formed in an orbit similar to the earth.

SINGER: I leave that to Urey. He has proposed the moon was formed separately.

ORÓ: Because it does not resemble the earth, but you are saying the moon formed close to the earth.

SINGER: I said it was formed in a similar orbit to the earth.

ORÓ: I suggest then it should have a composition similar to that of the earth.

SINGER: Sorry about that. You may assume that the moon lost the iron or acquired silicates (Singer and Bandermann 1970).

If anyone wants a write-up of this, it was published in the Geophysical Journal of the Royal Astronomical Society (Singer 1968). If you drop me a note, I will send it to you.

HOROWITZ: Thank you very much, Fred.

MICROBIOLOGY OF LUNAR SAMPLES

HOROWITZ: I asked Dick Young to summarize the microbiological work that has been done on the lunar samples.

YOUNG: This was a last-minute request so I don't have much data nor have I any slides.

Two laboratories have looked directly at the question of whether or not there are viable organisms present in lunar material. One of course is the Lunar Receiving Laboratory (LRL) at Houston and the other is the NASA Ames Research Center at Moffett Field, California. If there are any questions on detail I am sure Chuck Klein knows considerably more than I do about what was done at Ames and perhaps at Houston, too.

I can summarize the results of both of these labs very briefly. They were negative. The Houston group was charged primarily by an interagency agreement to inspect lunar material from the point of view of pathogenic organisms: Is there anything in lunar material that poses an epidemic threat to any form of terrestrial life? They arrived at a fairly complex protocol for testing a tremendous array of terrestrial organisms, far more than could be accommodated in the facilities of the Lunar Receiving Laboratories. This was ultimately reduced to a more limited spectrum of organisms. I don't have the complete list, but it included plants of interest to the Department of Agriculture. They looked at radishes, wheat, spinach, liverworts, and so forth. They studied many animals, including fish, insects, and whale eggs.

WASSERBURG: In the middle of the testing there came the great announcement that the female whales were laying very well.

HOROWITZ: Many tissue cultures were studied, too.

YOUNG: Yes. Two conclusions were drawn. There was no evidence that pathogenic organisms were present in any of these assays. The second conclusion was that there were effects on some of the test systems, as illustrated in this viewgraph (Fig. 21).

MARGULIS: What was the control?

YOUNG: The same plant grown in agar without lunar samples.

MARGULIS: Agar?

YOUNG: Grown on an agar substrate.

RICH: An agar substrate low in titanium?

YOUNG: Right.

HOROWITZ: They have tried to reproduce this with appropriate nutrients, I guess.

YOUNG: Yes, they are in the process of doing this. As far as I know, why liverworts grew faster is unexplained.

MARGULIS: They must test an array of substrates–not just plain agar. Since agar is inert, they must have included some salts.

YOUNG: Of course. Was it grown in soil or was it in agar substrate?

WASSERBURG: It was on an agar medium.

MARGULIS: The agar mineral content has to be different from the lunar.

YOUNG: Right. One control was the same medium to which lunar soil was added. Whether or not this was a trace element effect or not is not known yet, but certainly this is a possibility. Similar effects in other plants were observed in addition to enhanced growth rate–some plants had better greening of the leaves.

As I recall, something like a 10 percent reduction in quail-egg laying was reported but this was considered to be within the error of the technique. They found that some of the agar containing lunar sample, when resuspended, would not support growth of bacteria. This suggests that the lunar material is either bactericidal or bacteriostatic. It is also possible that the bacteria simply won't grow in an old medium, but this is still an unanswered question, too.

The Ames Laboratory, not restricting itself to a search for pathogens, looked for any evidence of growth of viable organisms. They used 50 grams of lunar material. They cultured 3,000 petri dishes using 9 media, varying both in composition and pH. These have all been described. I won't detail them. They used 3 gas mixtures and 4 temperatures with a total of 3,000 dishes, and found no evidence of viable organisms. In a sense, to culture 3,000 dishes without even having contaminant is remarkable.

This may be interpreted in an interesting way. In this evidence of bacteriostatic or bactericidal activity or something else.

RICH: Did they have controls that they inoculated to show that the media would support growth?

Figure 21. Effects of lunar material on seedling growth.

YOUNG: Yes. It was a very carefully controlled experiment.

RICH: Did you also put lunar material in with bacteria?

KLEIN: They are doing that now.

YOUNG: I think the question of whether lunar material is bactericidal or bacteriostatic hasn't been adequately answered yet. I found out about another relevant experiment just before I left: Maybe someone here knows more about it. Apparently organisms have been recovered from the Surveyor III spacecraft parts that were brought back, suggesting pretty high survivability.

YOUNG: How long was it up there?

WASSERBURG: Almost 3 years.

RICH: Which organisms?

YOUNG: It is known but I don't know.

STROMINGER: I thought everything sent to the moon was supposed to be sterile?

YOUNG: No, the spacecraft were clean but not sterile.

WASSERBURG: They brought back the camera system, the scoop and the analyzing equipment for bugs, in plastic insulation.

YOUNG: That's right, it was in plastic.

ORGEL: Were they supposed to be bugs or spores?

WASSERBURG: Teflon.

RICH: The bugs were buried inside the teflon?

WASSERBURG: That is what they are trying to find out. They are looking for organic things. It was a mixture of teflon and other not purified basic things.

MARGULIS: Out of these 3,000 plates, there wasn't one culture of any kind?

YOUNG: That is correct.

KLEIN: Not only that, but there was nothing growing on plates at Houston where they probably made up thousands.

MARGULIS: Did all of them have lunar material?

KLEIN: The 30,000 represented dishes with lunar material in them and I would say probably there were at least that many at Houston. Oyama (Oyama et al. 1970) has recently found that enough titanium and chromium is leached out on these plates locally to give very high concentrations of these ions around the specks of lunar material. These concentrations of ion ought to be toxic.

MARGULIS: If you set up these 3,000 plates only without lunar material, you would have at least a background noise of how much contamination is expected by chance?

KLEIN: Zero.

The two labs approached their work differently. The LRL at Houston was set up to contain all potential organisms from the lunar material. All the air-conditioning and everything was such as to keep the material there, perhaps contamination might be expected on those plates because air was going into the system.

Ames was set up with exactly the opposite strategy and viewpoint: to keep anything extraneous from getting in and contaminating, so our air system was just the opposite.

Neither laboratory found any organisms on any plates. It was absolutely fantastic.

RICH: Is this true for both Apollo 11 and 12?

KLEIN: Just 11.

YOUNG: They are just starting on 12.

MARGULIS: How much lunar material per plate was this?

KLEIN: In Oyama's case I believe he sprayed roughly 100 mg per plate, maybe 50 times.

RICH: You can't do that, can you? You have only 50,000 mg for 3,000 dishes.

KLEIN: Maybe he used 10 mg.

WASSERBURG: They had a hell of a lot of material.

YOUNG: In the paper it says 50 grams were used.

KLEIN: We also noted that the astronauts' suits were leaking microorganisms at the rate of 30,000 per minute per man. It is incredible that no contaminants were found.

We also know when the astronauts scooped up the material and threw it into the spacecraft, it wasn't hermetically sealed. I really don't understand why the plates showed nothing unless these ions or these metallic components were really very powerful. Let me make clear though that hundreds of animals got massive doses of lunar material: shots in the peritoneal cavity; the stuff was rubbed on the backs of turtles and the underside of plants, and God knows what. Nothing happened to any of the animals or plants, so it seems as though it is a rather specific microbial agent.

MARGULIS: How did they introduce the lunar material into the plates?

KLEIN: They sprayed it.

MARGULIS: And the control they just sprayed with nothing in it?

KLEIN: Yes.

WASSERBURG: It is very hard to believe that you could have an absolute zero blank. Maybe the lunar material is herbicidal or suicidal or that at some time you didn't get a bug that would grow in the agar plate when you just sprayed it on the straight controls–that is a little hard to swallow.

KLEIN: I believe the result is valid; they ran through several simulations and the system was designed to keep contamination out.

ORGEL: One wonders whether the system would show up normal bugs if they were there.

KLEIN: Of course it would, we have competent people.

WASSERBURG: But it is a matter where one would have expected something.

KLEIN: Perhaps this is an unrelated finding. The LRL, when they also set up a simulation, went into the backyard and ground up some rocks–T. R. Bell described how he did it. He picked it up in his hand, crunched it up in a mortar and pestle and it went through their system and they didn't find anything.

RICH: Rock from the backyard? Houston soil is not that sterile.

WASSERBURG: They washed it in peracetic acid before they put it in the system.

KLEIN: You said it; I didn't.

WASSERBURG: This doesn't make sense.

HOROWITZ: It makes sense if their medium wouldn't support life.

RICH: Did they wash the sample in acid first?

KLEIN: I doubt it. They weren't surprised. They were happy.

RICH: Did they think the soil was sterile?

ORGEL: They didn't sterilize the soil and then become happy because they had shown they had sterilized it?

YOUNG: They were happy because they were able to find one organism.

ORGEL: You mean one.

YOUNG: Yes.

HOROWITZ: I guess we are here to discuss facts, not miracles.

WASSERBURG: Actually, it is a serious business. I wonder if Dr. Young would comment about where the research report on the Apollo 11 will be published so that the tests are available to the scholarly community.

YOUNG: The Ames Report or the LRL Report?

WASSERBURG: Both.

YOUNG: The Ames Report is in *Science* (Oyama et al. 1970).

WASSERBURG: What about the LRL statement?

YOUNG: I believe the LRL statement will be published as a NASA special publication.

SHELESNYAK: We are asking questions that seem to suggest the feeling here that the people who did this work were completely incompetent. I don't think that is right. If you have the proper setup 3,000 plates can be run without contamination. Good laboratories do this. I don't think it is fantastic.

RICH: But what about taking a piece of soil and grinding it up?

SHELESNYAK: I don't know what he means and I would be reluctant to comment. With all due respect, we have a very poor description of this particular experiment.

KLEIN: It is very central, Shelly.

SHELESNYAK: Possibly the soil had antibiotics in it.

KLEIN: Yes, that's possible.

RICH: Antibiotics aren't usually that broad spectrum as far as soil organisms are concerned. They will knock out some but not all of them.

SHELESNYAK: We don't know how many plates were run and if he took the sample in his naked hand or in a glove.

KLEIN: Bare hands.

ORÓ: Then, I don't understand.

I don't understand why this experiment was done, and secondly, I don't understand the results.

SHELESNYAK: Why did someone go out and take a hunk of dirt with dirty hands and put it in a dirty mortar and pestle and throw it on a plate? That, I don't understand.

KLEIN: You would have to live at the Lunar Receiving Laboratory to understand. In the hectic days just before the Apollo 11 flight, something called a simulation was performed. Since there was no lunar material, they went out and got a rock, a couple of rocks. The objective was not to find organisms but test if the gas would leak, whether the vacuum would hold, whether the technicians would sweat, and so forth. The objective was merely to start on page 1 and end on page 50 of the so-called protocol. When they got all through, everybody was happy. The fact that they didn't find any organisms—

RICH: The object was to look for leaks, organisms coming from other sources. If that was the experiment they would have put it in peracetic acid; they would have tried to sterilize it.

KLEIN: They didn't. I followed this closely. I was on the ICBC and apparently was the only one that was worried about the fact that they couldn't find organisms in the soil.

MARGULIS: Didn't you say they found one?

KLEIN: They reported finding two colonies of one kind of organism.

ORGEL: How many plates?

KLEIN: They ran through the complete protocol. That means injecting the stuff into bunnies and God knows what all.

Let me also say on some of the lunar samples—not the backyard garden stuff—somebody reported finding evidence of peracetic acid and freon in the gas chromatograms. It is not implausible to suspect that samples we received at Ames had, in fact, been partially, briefly sterilized.

WASSERBURG: Yes. Calculation attempts were made. A few of us did some arithmetic on the peracetic acid. I don't think there was enough that one could calculate that an attempt had been made to really wipe out the whole box.

RICH: I see, so the possibility exists that the Apollo 11 biological results really aren't all that conclusive; that maybe they sterilized by mistake the whole box——

WASSERBURG: They didn't do that.

KLEIN: It would be important to know how much of it had been exposed to peracetic acid. I don't know that.

RICH: This sounds like a very strong argument that it should be done again.

KLEIN: It ought to be done right.

WASSERBURG: If you don't do it right, forget the Mickey Mouse. There is no sense in running through it and pretending you have done an experiment when you haven't done an experiment.

RICH: It ought to be done again correctly.

YOUNG: That recommendation has been made. I don't know whether or not it will be accepted.

WASSERBURG: The recommendation does not demand such stringencies as would upgrade the experiment and make it that meaningful. The recommendation does not represent a sense of what you would call a systems appraisal.

YOUNG: That wasn't the intent. The recommendation was simply to express dissatisfaction with what has gone before, which it did.

EDITOR'S NOTE: *Dr. Wasserburg then challenged "the biologists" (by whom he meant everyone not a geologist or astronomer) to collectively and wisely become involved in the decision-making processes related to the receipt and disposition of lunar materials for biological experiments. He then thanked everyone for attending.*

MARS

MARS
Sunday Morning Session

PROBLEMS OF RELEVANCE TO THE ORIGIN OF LIFE

The conference reconvened at 9:15 a.m. with Dr. Norman Horowitz presiding.

HOROWITZ: Today, as you know, Mars is on the program. The history of Martial studies has been an amazing history of error. We could go back and talk about Percival Lowell and his contributions to our misunderstandings about Mars, but we don't have to go back that far. For example, just 10 years ago it was widely believed that Mars had an 85 millibar atmosphere composed of over 90 percent nitrogen, that its polar caps were made of water ice, that the Martian maria contained organic matter, based on their reflection spectra. We can come up to the present day with misunderstandings. Even during the Mariner 6 and 7 missions, one principal investigator was convinced he had discovered life on Mars as recently as last August and at an unforgettable press conference went so far as to describe a rather detailed model of the Martian ecology.

These ideas have all turned out to be wrong and the strange thing about these errors has been that they all have been in the same direction. They have all tended to make Mars more earthlike than it really is. History must decide whether these errors are due to the inherent difficulty of studying Mars or whether there has been some kind of systematic self-deception on the part of Martian observers.

This brings me to the introduction of our discussion leader, Professor Bruce Murray, Professor of Planetary Science at Cal Tech. Bruce has been deeply involved in the Martian investigations and he has, more clearly than most, seen the necessity for stripping the romantic veil and seeing Mars as she really is. Bruce has done valiant duty and contributed greatly to the ongoing process of understanding Mars.

MURRAY: To get into the subject it might be interesting to list on the board the questions we collectively have about Mars that the recent data from spacecraft, the ground, or the orbiting astronomical observatory could be relevant to. This concern with origin of life and life on Mars is old and interesting, but I think we need to stand back and ask what it is about the planet that is knowable and in some way related to the origin of life question, even if quite indirectly. A minor example of this occurred yesterday when Dr. Oró referred to the need to understand the exposure ages of the Martian surface materials and the thickness of the regolith—if you want to use that word although it means something different on Mars than on the moon.

There is now some indirect information which is at least suggestive of differences from the moon. There are questions of water vapor and particularly of evidence of soil moisture. Before we start any presentation of Martian data I thought it might be useful to go around the room and have people cough up questions concerning the planet that may or may not be answerable by the rather large amount of data that has come in.

ORÓ: You want questions or comments?

MURRAY: Questions. Break your question down into specific questions that are knowable on the planet within this time frame. I will ask the speakers to try to make relevant their observations to this list of questions if they can, and in this way we will try to make sure we do focus on the biological relevance of the data and not have a seminar on Mars which would not be an efficient use of our time.

ORÓ: What is the thickness of the so-called regolith and the turnover rate? I am puzzled formulating this question because even though it bears a relation to the moon, I can't relate it to the earth.

MURRAY: The earth is a totally different sedimentary process. There is no analogue on the earth nor is there any analogue on the moon for the sedimentary processes on earth.

ORÓ: Then are we in a sense biasing our questioning?

MURRAY: Suppose I call this a *soil profile*.

ORÓ: No, *soil* has been used profusely by lunar experts because we haven't had someone like McClaren or a soil specialist to tell us not to use the word because there is no such thing as soil on the moon. Let's not talk about soil on Mars because we don't know.

MURRAY: Let's call it *debris mantle*. We have been trying to find the right term among ourselves. We mean the aggregated, the broken up material that at some point is underlain by consolidated material sometimes called bedrock. There are a variety of processes that have to do with the mixing and thickness of that.

Toby, what describable aspect of the Martian picture do you think has relevance to organic material or planetary life?

OWEN: I am biased by my own interests, but I think the water balance problem is relevant.

MURRAY: By *water balance* do you mean liquid, solid and vapor H_2O all together, the water cycle?

OWEN: Yes.

ORGEL: What is the reliability of the assessments of the amounts of atmospheric and surface nitrogen?

MURRAY: I'll list *nitrogen cycle* to keep the terminology consistent.

ORGEL: If there is nitrogen, what is the oxidation state likely to be?

MURRAY: O. K. The question of oxidation state and the role of oxygen is another question.

RICH: I think it is a mistake to approach the problem this way, trying to second guess the nature of life on Mars. I think it blurs the fact that you, who study Mars, have hard facts, soft facts, and very soft facts. We non-Martian specialists and non-planetary astronomers will blur in our minds that there are certain hard facts about Mars that we should know and soft facts we should know as soft. Now we will only give you obvious queries based on our concepts of terrestrial life. Carbon and temperature are important of course, but this is not the best way to approach the discussion; it is too geocentric. There may be our type of life or no life on Mars, there may be a different kind. We don't know. We now ought to talk directly about the information available and its degree of hardness.

MURRAY: But the information far exceeds the capacity to assimilate it.

MARGULIS: He simply wants to organize the information around the relevant questions.

MURRAY: I want to discipline the speakers, whatever they talk about; it should be relevant to the issues. I really can't prejudge the experiments. Remember we—the United States collective taxpayers—have paid $120 million for Mariner 4 and $150 million for Mariner 6 and 7. That is more money than has been spent on entire fields of science. If we can't even organize the problem of search for life, I don't think we have a very good selling point. We must organize our search in observational terms, all we can do is observe a great deal.

It is only one strategy but we have all already begun to spend another $150 million on Mariner '71. So we want to be sure we are studying the problem. Maybe I am reacting a little strongly but this problem, to me, as a nonbiologist interested in Mars, has frankly involved too much soft thinking and uncritical acceptance by non-specialists of marginal information. This is what Norm implied. We must ask, What experimental observations are relevant? To say, for example, "Look for cases," is not very helpful. We can't look for cases even in this context of this $150 million shot. I am trying to some extent to force this group, including the speakers, to come to terms at the boundary of relevant observations for the problem of life.

RICH: The list is straightforward. You are interested clearly in temperature profiles and the carbon cycle.

MURRAY: Temperature profiles, no. Everything we know now about temperature was predicted and is completely determined by solar insulation and radiation. There is no independent information on the polar cap, and the nature of the volatile phase. It would have been astonishing if there had really been departures from predicted surface temperatures.

O. K. We need carbon cycle and we know some about CO_2 and CO. Stanley Miller has thought about some phase relationships that may be involved. There is still only negative information about organics. Stanley, what do you want to add?

MILLER: I've been sort of preempted. I think nitrogen may even be more important than the water. We also might discuss if we can expect any relevant information to the presence of life out of the '71 Orbiter.

SCHOPF: I would be interested in learning about the wave of darkening that comes across the Martian surface apparently–

MURRAY: Would you settle for the nature of light and dark markings?

SCHOPF: Yes. I'd also like to discuss the geologic differentiation of Mars, and its history, the evolution of the planet, and its atmosphere and hydrosphere over time. I suspect that involves much speculation.

MURRAY: No. I can argue the reverse. We know more about that than much else now. There is an argument that Mars never had an ocean or atmosphere.

OWEN: It has an atmosphere now.

MURRAY: No, it lacks a terrestrial aqueous atmosphere which we presumably are worried about.

OWEN: I don't think it is entirely settled.

MURRAY: O. K., we can learn about this. The problem of differentiation and atmospheres go together because the only information we really have is from the surface.

ORÓ: Give us 25 percent and we perhaps shouldn't go to Mars.

MURRAY: There are others who have wondered if maybe some redistribution of priorities might be useful.

Conway, you are of course an observationalist, but do you have your own questions?

LEOVY: Mine have already been covered. I am interested in the interaction of the atmosphere with the surface, modifications of the surface through erosion or other processes, and exchanges of materials such as water vapor at the surface. I am generally interested in the atmospheric environment.

KLEIN: I think that in this interesting exercise we are going to come out with the same outline that we were given 2 weeks ago. I would like to know something about the current thinking on surface radiation; solar protons, x-rays, and so forth.

MURRAY: There are some pretty good data about this question. Unfortunately Charlie Barth, whose data is probably the most relevant, is not here but Conway probably knows it well enough to cover it.

Larry?

SODERBLOM: Related to the question of the light and dark markings, I think it is important to discuss transient phenomena. I don't necessarily mean the rate of darkening that Brad Smith finds and thinks is understood, but for example, the transient phenomena on Hellas, and so forth, that indicate that poorly understood dynamic processes currently occur on the surface. When we say primordial surface, we should include the question of its activity.

MURRAY: It is useful to make this distinction between general transitory features and the light and dark markings. Light and dark are reasonably well described and to some extent predictable. The question of what the "wave of darkening" really is and how much it is a statistical sample of this is still debatable. If Brad Smith were here I believe he would argue that he can't be sure there is a "wave of darkening"; but there are some correlations like that but not necessarily a uniform occurrence, for example, the varying brightness of Hellas.

SCHOPF: Are there any data suggesting local geothermal areas which would be relevant to this matter of internal history, and differentiation of Mars?

MURRAY: Not really. The only way these might be detected is by observing thermal anomalies, but we couldn't possibly resolve them with present techniques.

HOROWITZ: If water comes out, you could see it.

MURRAY: Not from direct measurement of thermal anomalies. From the albedo on Mars we can calculate better than we can measure the surface temperature. Now an anomalous appearance of a frost or white cloud at some point might be interesting. I am trying to force you to translate the question into something measurable.

Relevant measurements. If it implies a measurement that can't be made for 10 years, it is not relevant to this discussion. We can't rule out the possibility of an oasis, but the question is how we could get to see it. An anomalous frost and/or cloud phenomenon, locally derived, staying or recurring in the same place might be suggestive of water moving in and out of that place in the soil. All I can do is list these kinds of questions because we can only observe *observables*.

ORGEL: Have you included the composition of surface materials?

HOROWITZ: As deduced, of course, from spectral readings.

MARGULIS: This question is probably impossible but I would like an estimate of how much liquid water there could possibly have been over the entire history.

MURRAY: The only way we can answer that is by surface history and by the water cycle itself.

MARGULIS: Can you put an upper limit on the amount of open water?

MURRAY: I have a very prognostic view of that which says there never has been any.

MARGULIS: How hard is your statement?

MURRAY: This particular topic will be forcefully exposed on both sides.

MARGULIS: How is carbon distributed? Historically, how has carbon been distributed? What is its concentration and what is its oxidation state? Could hydrocarbons ever have been associated with any water?

MURRAY: My answer again is negative. There are no relevant observations.

MARGULIS: How much agreement is there on the negative statement?

SODERBLOM: We must begin with assumptions of original sources of carbon and so on and then we can talk about the evolution of surface processes which redistribute and modify surface morphology. But since everything depends on our assumptions of what the sources of carbon are, this is a problem.

HOROWITZ: Everything I am interested in has already been mentioned.

SHELESNYAK: Being on the bottom of an erudite deck like this, I lack both the wisdom and the wildness to state a question.

MURRAY: Sometimes that requires a strategic blend. Dick?

YOUNG: We shouldn't restrict our discussion to the question of life on Mars. Mars to me is of less interest as an object of study if there is no life, but Mars is an object of equal interest to the question of the origin of life, whether or not life is there.

MURRAY: Mars is of equal interest but not from the standpoint of the distribution of funds. For a tenth of the money proposed for a Lander program, we can find out what Uranus and the satellites of Jupiter are like. I do believe in equal partition as a function of ignorance.

ORÓ: But before going to the outer planets we must study at least one of the inner planets.

YOUNG: We need a comparison for the Earth.

MURRAY: It is difficult to disentangle these feelings. In 1960 when many people thought Mars was most likely to be the twin of the earth, we set different priorities. Now Venus is probably more like a twin and Mars is its own planet, not like something else.

YOUNG: My question is why life on Mars didn't happen there as on earth.

MURRAY: Then let's put down significant negative results whenever they can actually be deduced negatively, as a way of forcing the question of why life didn't develop.

The question we should talk about after three or four cocktails—if you accept for argument's sake my pessimistic view of life on Mars—is what else can we do within our lifetimes with our technology relevant to the question of the origin of life? That is the real question to be examined by this group. Should we build a kilometer radio telescope in earth orbit and listen to the signals from other stars? As the probability for life on Mars now drops, there certainly must be other candidates that ought to be looked at. This is different from not caring about life, it insists on proper perspective.

RICH: Do you really want to discuss priorities in the sense of it we shouldn't go to Mars, what else should we do?

MURRAY: My personal suggestion is that after we finish the discussion on Mars and we accept the fact that the *a priori* probability of finding life there is very low, and we still give the same absolute priorities to trying to get information about life outside the earth in some form, we should consider what to do about it. What is the next direction to go?

YOUNG: Another point of view is that we are trying to understand planets, and we have got this earth that is all messed up with biology. How does a planet not messed up by life evolve?

MURRAY: To counter this argument, I will prove to you beyond reasonable doubt that the probability of life on Mercury is the same order as the probability of life on Mars. Therefore within a factor of 10 we ought to be spending the same kind of resources. You are saying a planet must be explored in order to search for life even if it turns out the probability of learning about life gets lower and lower. Your argument is different than that of a planetologist. It is my profession to study planets, dead or alive or in between. This group must ask, What about the search for extraterrestrial life? Mars is a subset here, but there are other possible options.

OWEN: I think the point Dick is making is that it is useful to study Mars in a way quite different from the study of Mercury, by trying to find out why there is no life on Mars, because of the insight we'd get into planetary evolution. Completely aside from the biological question, planetologically, Mars is a very different object and in a different class from Venus.

MURRAY: But this is a conference on the origin of life.

ORÓ: I don't know whether you know the origin of the Viking Project but I infer basically that it came from the meetings at Palo Alto on exobiology. There Dick Young's point of view was influenced by several people, including myself, who said the only way to understand any unique process like life in the solar system is by understanding the solar system as much as you can. Equally important to finding a planet with life is finding one that has no life but is on what you may logically conceive are the earliest steps toward making it. I think this is the point Dick is trying to make. At Palo Alto, I said this is the only way the system cannot fail.

MURRAY: The judgement that Mars is indeed a suitable planet for that test is based upon decade-old information. It is a little like looking for the dime under the light. The drunk was down there on the ground trying to find the dime that he dropped and the policeman came along. Offering to help, he said, "Where did you lose it?"

"Over there."

"Why are you looking here?" The drunk answered, "There is more light here."

This may be an act of desperation.

ORÓ: Your judgment is very subjective right now.

MURRAY: But less subjective than it was a decade ago.

ORÓ: I grant that.

HOROWITZ: Let's talk about Mars.

MURRAY: For most of the day, we will talk about relevant processes that can be inferred or are at least consistent with the observations. That is all we can do.

STROMINGER: Are microscopic observations included in this?

MARGULIS: Do you have any information about phosphorus and sulphur.

MURRAY: Only negative–there is no SO_2.

ORÓ: This would be part of surface composition.

MARTIAN ATMOSPHERE FROM GROUND-BASED OBSERVATIONS

MURRAY: We have decided to take the morning to discuss the low density volative question: the polar cap, the atmospheric composition, the blue haze, observations on water vapor and water cycle. Conway Leovy will try to organize the discussion from the point of view of the Mariner 7 results and Toby Owen from the point of viewof the ground-based observations. In the afternoon we will consider solid body questions, surface history, internal pressure, and so forth. Toby, will you perhaps begin with facts deducible from the ground-based observations?

OWEN: After this talk about hard and soft facts I am reluctant to present any facts at all.

I thought I'd tell you about the currently popular numbers and give you my opinion about which are harder and which are softer. Although I need not go into the techniques here in detail, I will be talking about results based on observations made with telescopes and spectrographs.

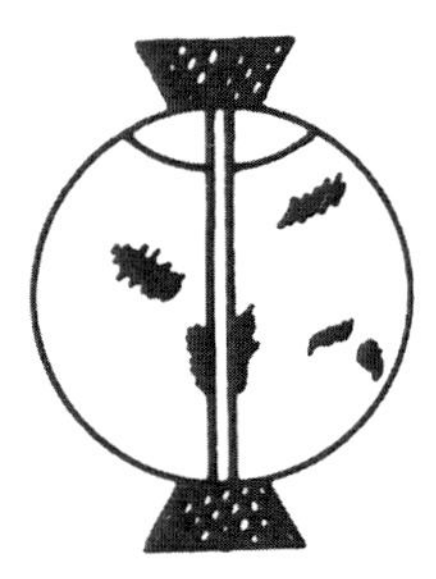

Figure 22. Upper: The appearance of Mars on the jaws of a coudé spectrograph slit.

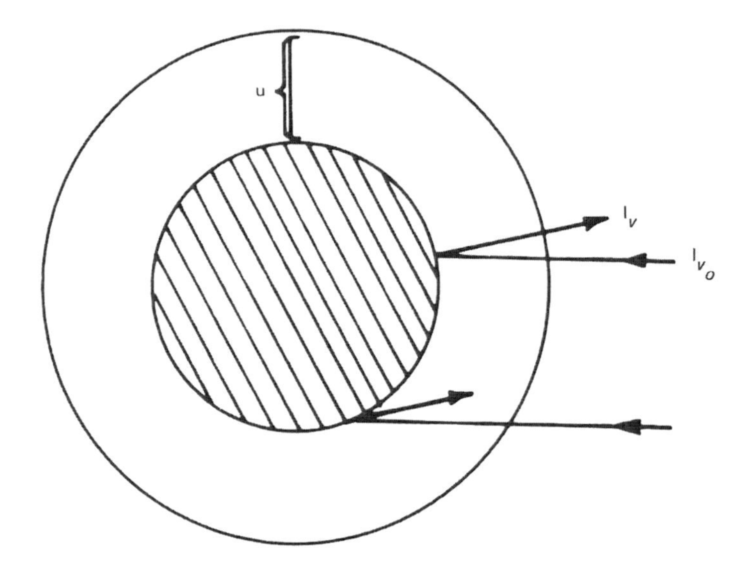

Figure 22. Lower: Incident sunlight (Iv_0 *passing through the atmosphere of a planet and being reflected by the surface. It emerges from the atmosphere as Iv, carrying information about the composition of both the atmosphere and the surface.)*

Figure 22 indicates how these observations are made. The lower sketch shows the planet and its atmosphere, with incident sunlight passing through the atmosphere and being reflected from the surface; it thus has a double path through the atmosphere. The spectrum we observe therefore contains much solar information in addition to planetary atmospheric absorptions– the light of course comes through the earth's atmosphere, too, which adds spectral lines that must be accounted for in the analyses.

The upper sketch in Figure 22 indicates what the image of the planet looks like on the slit of the spectrograph. If there is a polar cap here, maybe a dark area and some light area, the width of the resulting spectrum contains some areal information. If the image is large enough it is possible to tell whether or not the carbon dioxide absorption varies over the dark and light areas, whether water vapor is concentrated over the polar cap, and so forth.

The problem is that the earth's atmosphere blurs the image and moves it around. To get these "soft numbers" for abundances of atmospheric constituents, you must integrate over a certain area of the planet because the exposure takes a rather long time. When you are told about a real resolution on a disk from some ground-based spectroscopy, you must be a little sceptical. That is my personal opinion. Such measurements are often not as good as they are claimed to be.

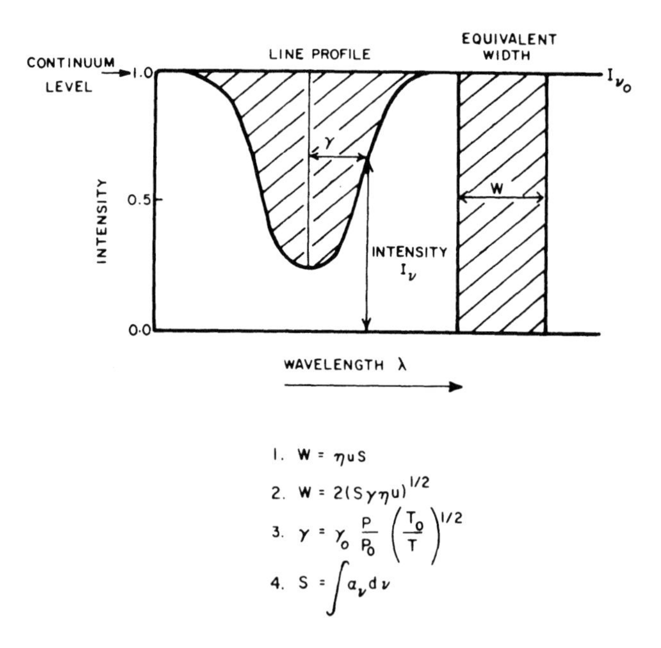

Figure 23. Appearance of an absorption line showing the various characteristic parameters and the ways in which they are related to the abundance (u) the pressure (p) and the temperature (T) of an atmospheric constituent.

Figure 23 shows a very schematic sketch of an absorption line showing that we can get information on the identification of an absorbing constituent from the position of the line, the wave length. From the area of the absorption line we can learn about the amount of the absorber. If the line is quite strong we can even get atmospheric-pressure information. With a whole series of lines, we can get a rotational temperature, a mean temperature in the atmosphere using this technique.

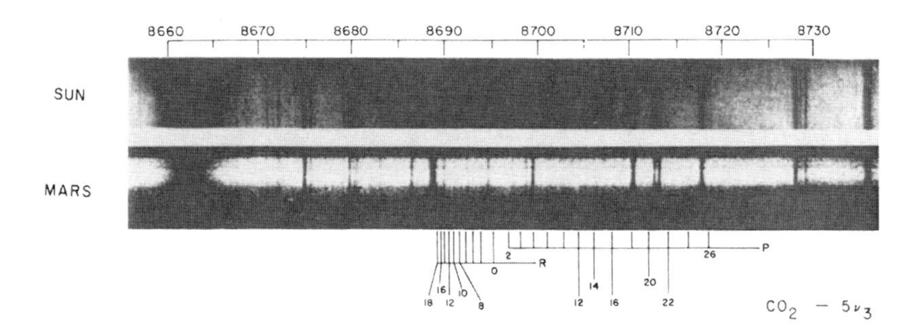

Figure 24. The appearance of Martian and solar spectra in the neighborhood of 8700 Å. Weak CO_2 lines appear only in the spectrum of Mars, indicating that they are contributed by the atmosphere of that planet.

Figure 24 shows some Mars spectra. This solar spectrum shown for comparison indicates what is coming into the planet. There are absorption lines due to elements in the sun, for example this iron line at 8689 Å.

After passing through the Mars atmosphere, new absorptions, not present in the solar spectrum, are added; they are carbon dioxide lines in this particular region. The absorptions are really very small.

Figure 24 shows a very weak band of carbon dioxide, one of the bands used to determine the total abundance of CO_2.

RICH: Is this from a ground-based telescope?

OWEN: Yes, so we are looking through the earth's atmosphere, including of course some CO_2 in the earth's atmosphere. This band is so weak that there is no absorption from the earth's atmosphere of CO_2, so we don't have to worry. When we look at some of the stronger bands it is a problem and a correction must be made.

LEOVY: Are those individual resolved lines?

OWEN: Yes. These are rotational lines. The problem is very messy when you are trying to find indications of water on Mars because even the weak bands of water vapor are quite strong in the earth's atmosphere. Water vapor measurements are best made when the relative motions of the earth and Mars are great enough that there is a Doppler shift resulting in the movement of the Mars line out from behind the telluric lines where they can be seen.

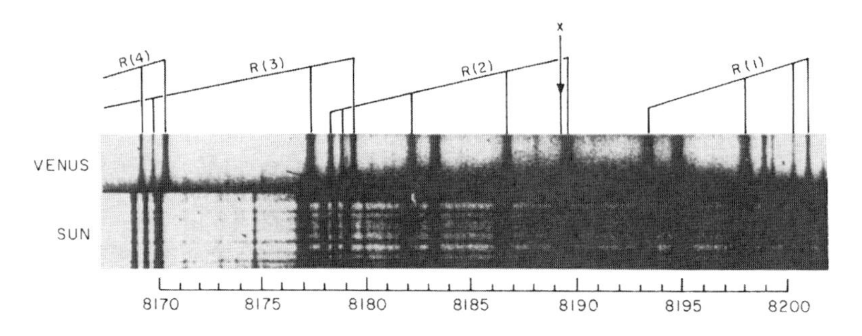

Figure 25. Spectra of Venus and the sun in the region of the 8200 Å water vapor band. Only one possible line is contributed by the Venus atmosphere (x); this seems likely to be a plate defect.

Figure 25 shows Venus, to give you an idea of the technique, in the region of a moderately strong water vapor band. Looking directly at the sun through the earth's atmosphere, the spectrum we get contains solar lines and water lines. The vertical lines drawn above the spectra indicate water lines. In this spectrum I have lined up some solar lines in the two cases. The solar lines reflected from the planet are actually shifted with respect to the direct spectrum of the sun, but by lining them up in this way (Fig. 24) the water lines in the atmosphere of Venus, if they were present, would appear above the water lines in the earth's spectrum shown below.

On Venus no water lines in this region are seen. The one faint feature (Fig. 25) probably is some kind of a plate defect. If there were much water present in the atmosphere of Venus, we would expect that each of these telluric lines would have a Venus component above it.

Figure 26 shows exactly the same region of the spectrum for Mars. These weak planetary components are here in each case; Figure 25 shows only a small part of a band that goes 'way on. Here every strong water vapor line does have a Martian component, indicating there really is water on Mars, much more than on Venus.

Let me write down a few numbers. The generally agreed upon abundance of carbon dioxide at the moment is 80 ± 15 meter atmospheres. A meter atmosphere—the number of molecules per square centimeter in a vertical column–is a convenient unit. One centimeter atmosphere is 2.687 x 10^{19} molecules per square centimeter.

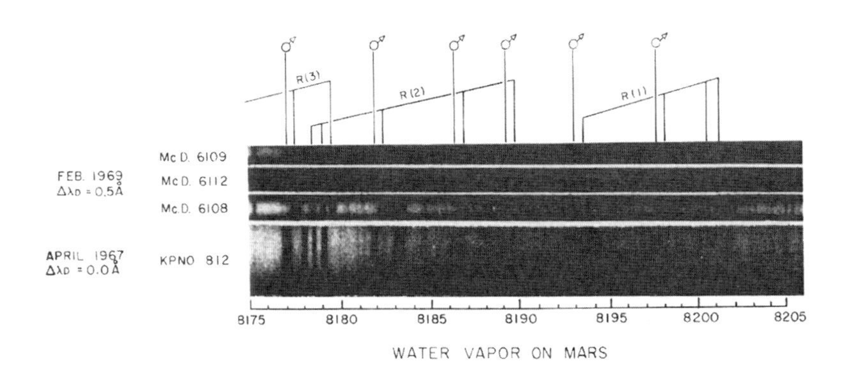

Figure 26. Spectra of Mars in the region of the 8200 Å water vapor line. The lowest spectrum was recorded at opposition and thus exhibits no Doppler shift. The Martian water vapor lines in the upper spectra are designated by the usual symbol for the planet.

Carbon monoxide has been found; it is about 5 ± 2 centimeter atmospheres–very much less. Oxygen was reported but even the investigators who reported it are not convinced they saw it. They give an amount less than or equal to 20 centimeter atmospheres.

Water seems to vary from 15 to 30 precipitable microns. This means that if all the water was condensed out it would produce a layer between 15 and 30 microns thick.

ORGEL: Can this be translated for those of us who don't know?

SODERBLOM: Put them in partial pressures.

OWEN: This is about 3.6 centimeter atmospheres of H_2O.

From spectroscopic investigation we also get the total atmospheric pressure. Various pretty soft numbers float around at the moment, but for the sake of a nice average the total atmospheric pressure comes out to 6.5 millibars with an uncertainty of about 2.5.

LEOVY: Are those limits based on the spectroscopic evidence alone?

OWEN: Yes.

SCHOPF: Is there some particular reason that oxygen is less easily detected spectroscopically?

OWEN: No, it is a very good, strong absorber. The problem is that the only people who claim they have seen it found two lines, one not even in a correct position and they had to invoke a rather peculiar pressure shift to account for the change. I would like to say that even whether there is any oxygen there at all is in doubt.

MURRAY: Isn't there an inconsistency because that amount of oxygen would imply a considerable amount of ozone and ozone wasn't seen in the ultraviolet?

OWEN: No, it is not really inconsistent with the ozone.

HOROWITZ: If there is CO there has to be some oxygen.

OWEN: Unless it immediately reacts.

MURRAY: There is some partial pressure.

HOROWITZ: There is a steady state concentration of oxygen in the atmosphere. It could be low.

MILLER: Could you convert the water into millibars and into dew point?

OWEN: I believe–if it is 30 microns I think it comes out to 200° Kelvin as the dew point saturation.

HOROWITZ: That is right.

MURRAY: The range is 190 to 200 depending on the number used.

OWEN: It is really very unfortunate that the uncertainty is as large as it is. It is very interesting that when all these components are added up the pressure comes out to be about 6 millibars. Is there room for a substantial amount of other gaseous components to give the total surface pressure? If there is, what would they be?

MURRAY: When you say "surface pressure" to what surface are you referring?

OWEN: The mean surface over the planet. The spectroscopy integrates over the whole disk.

MURRAY: The problem is the topography on Mars is large.

OWEN: There are 12 kilometer differences.

MURRAY: And the scale height is of the order of 10, so the topography is of the order of the scale height. Consequently the pressure, indeed as on the earth, must be specified at the altitude at which there is fadeout. This affects spacecraft measurements very significantly.

OWEN: Yes. I want to mention that, but my point here is to talk about other possible atmospheric constituents. Of course everyone's old favorite is N_2 but we can't say anything about nitrogen from the ground-based spectroscopy because it lacks absorbing lines in the region in which we could directly detect it. This is also true of argon, another very good candidate.

As Bruce mentioned, there are large differences in topography, on the order of 12 to 15 kilometers, as detected by the radar. If we are talking about a mean level here we still might have pressures of 15 mb. in the low areas. This would not really be out of the question, especially when you consider that the radar is integrating over a 300 kilometer area, which is pretty crude.

But we expect–except for the water vapor–that all the atmospheric constituents are uniformly mixed; therefore the problem of whether or not there can be substantial amounts of nitrogen or argon remains.

Some attempts to do topographic mapping, looking for these irregularities via the carbon dioxide band have been made. Personally I think variations in the CO_2 absorption are too hard to measure and radar should be relied on.

MURRAY: But the radar profiles do not appear to be looking at topography smaller than their size, so there is no evidence from the radar that they are indeed sampling a rough surface below that size. The evidence from the radar is that the height features are large scale in dimension.

OWEN: We should have much better information about that from the '71 Orbiter.

Of course the water vapor varies. I'm still not sure we have been seeing really measurable differences, from year to year. The group–Ron and Schorn and Bob Schultz at the MacDonald Observatory in Texas–this year has found–with adequate resolution for the first time, I think–that the water does seem to decrease as the north polar cap grows. Their initial observations made in early summer in the northern hemisphere of Mars indicated that there was more water in the northern than the southern hemisphere of the planet. Although there do seem to be seasonal variations in the water vapor abundance, we need more information.

An additional observational problem is that in order to get the large Doppler shift needed in order to detect the water vapor line, the planet is observed when it is very far from the earth. This means a very small image and unusual difficulty in obtaining good spatial resolution. This is something that would be very nice to work on from the '71 Orbiter observations.

This has all been ground-based observations. We also have new results, shown in Fig. 27, from the orbiting astronomical observatory–OAO–which permit us to look in the region of two to three thousand angstroms. This shows the ground-based results down to the ozone cutoff; where at the arrow [pointing] the OAO takes over. These two curves show two different phase angles of the planet. Here is the rise toward shorter wave length expected from Rayleigh scattering; the smooth line drawn through the crosses is for an atmosphere of 80 meter atmospheres of CO_2 at a surface pressure of 6.5 millibars. It fits the data very nicely.

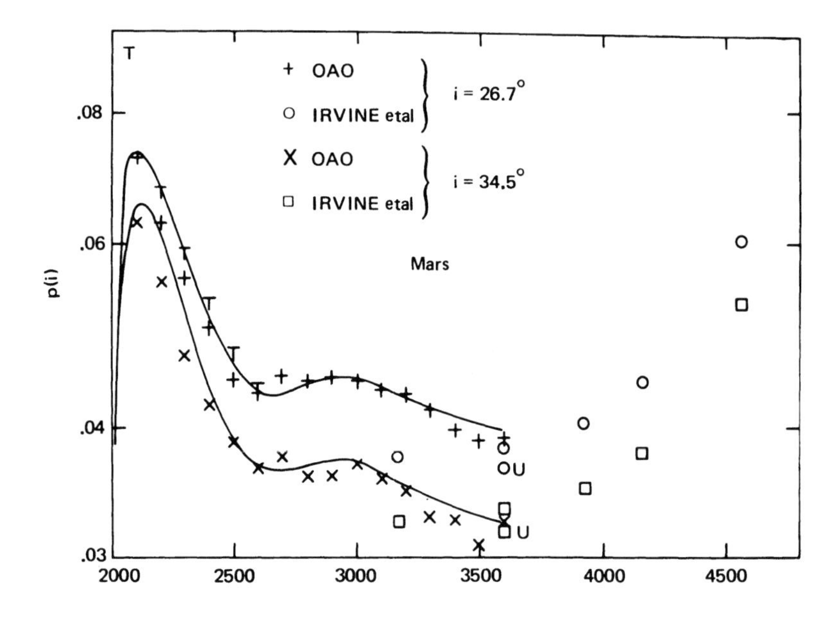

Figure 27. OAO spectra of Mars (preliminary). The rise in intensity toward shorter wavelengths predicted by Rayleigh scattering appears to be interrupted by an absorption at 2600 Å.

The only anomaly is this funny dip at 2500 angstroms. It has been matched in models by assuming the presence of a small amount of ozone. We who work with these data are still not sure whether we really are seeing an atmospheric absorption here. When the reflected solar spectrum is divided out there is the problem of strongly concentrated solar absorption lines right at this wave length; a slight displacement in the divided spectrum may lead to funny effects.

Charlie Barth (Barth 1970) suggests that here we are looking at the difference in albedo between the polar cap and the rest of the planet, and since we are integrating over the whole planet we get this funny effect.

MURRAY: The flat portion from 2600 on out is reflected as surface reflectance and the deeply rising part is at the surface reflectance, is that it?

In other words, that line with the 4 dependence goes zilch down by about 2600-2700 Å?

OWEN: Yes, especially in a small amount of atmosphere.

HOROWITZ: Why does it drop off at 2000Å?

OWEN: That is where the CO_2 absorption is coming from.

HOROWITZ: Are these data published?

OWEN: No.

ORGEL: How are the curves—and the dips—fitted to the lines?

OWEN: This is reflectivity and the curves are based on a Rayleigh scattering atmosphere. The dips include a little ozone absorption. The ozone is in the model. It could be a surface absorption also. We can't really say.

SCHOPF: Without the ozone, the line would be smooth?

MURRAY: Yes, of course, there are minerals that have strong absorption bands. A lot of silicon absorption starts at around 2600 Å too, to complicate the problem.

OWEN: Yes, and I am sorry to report that C_3O_2 absorbs here and, in fact, imitates the ozone profile beautifully.

Now, from the fact that this matches the 80 atmospheres so nicely, we can draw some conclusions about gases which are not present in the atmosphere. Figure 28 gives some preliminary upper limits in parts per million.

MURRAY: What does WEP mean?

OWEN: That is the Wisconsin Experiment Package. We have to acknowledge our sponsor the first time around. These are components we don't observe, and you can see that the detectivity is really very sensitive, although with respect to the nitrogen oxides we can do a little better from the ground.

ORGEL: Is CO_2 taken as 100 percent?

OWEN: Yes. These are very conservative limits. Carl Sagan* would like to drop everything at least by a factor of 5 or so, but I think they are reasonable.

MURRAY: Is the limit from the ground-based spectroscopy better than that?

OWEN: I doubt it. The great thing about the UV is that all these molecules have fantastic absorption rates.

**Center for Radiophysics and Space Research, Cornell University, Ithaca, N.Y.*

UPPER LIMITS ON MINOR CONSTITUTENTS IN THE MARTIAN ATMOSPHERE FROM OAO-A2

T. Owen

Hydrogen Sulfide	(H_2S)	< 2 ppm
Sulfur Dioxide	(SO_2)	< 2 ppm
Ammonia	(NH_3)	< 2 ppm
Nitrogen Tetroxide	(N_2O_4)	< 5 ppm
Nitrogen Dioxide	(NO_2)	< 10 ppm
Nitric Oxide	(NO)	< 10 ppm
Ozone	(O_3)	≤ 0.025 ppm

Notes:
1. From visible region of spectrum $NO_2 < 0.01$ ppm
2. 1 ppm = 8×10^{-3} cm atm (assumption is 100% CO_2)
3. HCN doesn't absorb in this region
4. C_3O_2 imitates O_3

Figure 28. Preliminary upper limits on possible constituents of the Martian atmosphere, derived from the OAO observations.

MURRAY: There has been some suspicion among the uninitiated that there was a problem with the photometry of the ultraviolet in the OAO.

OWEN: There is. The solar spectrum itself turns out to be rather poorly defined below 3,000 angstroms. Efforts are being made to define it in more detail, but this has caused some problem.

This obviously will also affect Barth's Mariner results, too, because we are stuck with the same polar spectrum.

MARGULIS: If you were forced to add the upper limit for N_2 to your list what would it be?

OWEN: I can't say anything about N_2, but at a meeting in San Francisco between Barth, Delgarno, and McElroy I believe the conclusion based on the Mariner results is that nitrogen is less than 4 percent.

ORGEL: How much does that depend on direct observation and how much on models of the atmosphere?

MURRAY: Can we postpone that important question and discuss it later when Conway talks.

OWEN: I'd like to end with a few comments about atmospheric evolution. As Norm said 10 years ago everybody thought that the atmosphere of Mars was predominantly nitrogen—because there was nitrogen on the earth and the CO_2 abundance was thought to be very small, about 2 meter atmospheres instead of 80. When it appeared that the atmosphere was predominantly CO_2 there was some initial dismay but it quickly became apparent that concise argument with data that Rubey had accumulated for the outgassing of the earth could be made to produce just this kind of picture.

If on this earth analogue model it is assumed that all the CO_2 that has been outgassed appears in the atmosphere, some conclusions about the expected amount of nitrogen and water can be drawn. Quite a large amount of water, much more than we see in the atmosphere and more than we expect to see in the polar cap, is predicted.

Of course, one possible place for it would be as permafrost. A vertical column of about 7 meters of H_2O is predicted on the earth analogue model and with the assumption that all the water has stayed on the planet.

MURRAY: That is also by comparison with the same 3 or so kilometers of water on the earth.

OWEN: Yes, this is comparative.

MURRAY: This is based upon estimating the CO_2 in limestones on the earth. The nitrogen is small as compared to the CO_2 on the earth—forget about it on Mars—and the scale of water to limestone—

ORÓ: Who has done this work?

OWEN: It was originally done by William Rubey (Rubey 1951) and is being revised by H. D. Holland (Holland 1964). The revisions are being made with some stimulus from the fact that there is so much CO_2 on Venus. The question raised is whether you can account for 90 atmospheres of pressure of CO_2 on the earth. I guess the answer is maybe you can if all the limestone is put back into the atmosphere.

ORÓ: Are all the so-called igneous carbonatites taken into consideration in these estimates?

OWEN: I don't know.

HOROWITZ: The what?

ORÓ: The carbonatites–a kind of carbonate different from the so-called sedimentary limestones. They are quite massive.

MILLER: These are always taken into account in estimates of the amount of CO_2. Unfortunately, the estimates aren't all that accurate but I believe they are within about a factor of 2.

OWEN: It is interesting that if you pursue this analogy, the amount of nitrogen expected is in the range of 2 to 4 percent of the CO_2. Within present detection limits this all seems pretty reasonable.

HOROWITZ: Assuming all the CO_2 is in the atmosphere.

OWEN: Yes, and also assuming everything has stayed on the planet.

HOROWITZ: Is this equivalent to saying there has never been much liquid water on Mars?

MURRAY: Yes.

MILLER: Is 7 meters considered very much water?

HOROWITZ: No. In order to arrive at this figure, to make the calculation, it is assumed that all of the CO_2 that Mars has ever outgassed is now in the atmosphere. Now that is multiplied by a factor of 15 or 20 and 3 meters of H_2O is arrived at. But if there ever were 3 meters of liquid water on Mars a lot of carbonates would have been deposited. In other words if there ever were an ocean, you can't assume that all the CO_2 is in the atmosphere now.

MURRAY: But conversely you can say what I personally believe, that there is more than 7 meters of water in permafrost because it is below freezing at a kilometer's depth below the surface of Mars. Much more than 7 meters can be stored around the planet; the CO_2 is not just the amount seen but quite a bit must be in solid form, too. It would be an extra-ordinary accident to be right at the transition point, and assume the nitrogen is there too–although some may have been lost or may be tied up with surface materials as well.

Present observations are consistent with the idea that Mars never volatilized much more than it did, the water never reached the surface but remained primarily as ice in some form.

HOROWITZ: The nitrogen value indicates that Mars never has had oceans of terrestrial magnitude. That is, if Mars is like the earth in these ratios, the earth would have 5 percent nitrogen in its atmosphere. If there had never been an ocean on the earth, and all the water had been ice, then practically all the CO_2 would still be in the atmosphere. The limestones would be in the atmosphere and we would have 20-30 atmospheres of CO_2 and 0.8 atmospheres of nitrogen.

MURRAY: Which, indeed, is what it looks like on Venus.

OWEN: Let me stress that the only number we know is the CO_2 value. Nitrogen hasn't been detected and we don't know how much water there is, so this is just based on the earth model. As Bruce has said, this suggestion is consistent but it certainly doesn't mean that is what has happened there.

ORGEL: This may be a totally unanswerable question but I fail to see the connection between the observation and these hypotheses. Naturally they are consistent, but can one conceive of cases in which the hypotheses would be inconsistent?

OWEN: Yes, I am coming to that. We have discussed the earth analogue situation, but let us think about other possibilities. We already know that the moon and the earth are very different, and it is presumptuous to assume Mars is the same as the earth. One other possibility is that a substantial contribution to the atmospheric composition may have been made by impacting meteorites and comets. The total mass of the Mars atmosphere is 10^{19} grams; this is about the equivalent of a large comet. So with a few cometary impacts at least some water and probably some CO_2 can be added to the atmosphere. That is one possibility.

MURRAY: A comet doesn't have an atmosphere.

OWEN: There is something called gravity. That makes the difference.

Another possibility even more remote, but perhaps not entirely out of the question, is that Mars outgassed more than what we are now seeing, but that some of its atmosphere was lost. Present ideas for planetary formation rely on magnetic fields to segregate the material. The original magnetic field may have remained for about a billion years, until the planet heated up enough to lose it. During that initial billion years, outgassing may have occurred and a lot of stuff may have come off. When the magnetic field was lost the atmosphere was subjected to solar wind bombardment which may have resulted in the loss of constituents.

McElroy and Cloutier and some of their associates have developed a model for the Mars ionosphere, involving the interaction of the solar wind, that seems consistent. The problem is to figure how much loss would occur and what would be lost, but at this stage of our ignorance I think we have to keep alternatives like this in mind. If the planet heated up, lost its magnetic field and much atmosphere, all bets are off and we really don't know what the total amount of volatiles outgassed was.

Bruce, later, will describe evidence of an episodic history of Martian surface erosion, one interpretation of which may be that Mars had a denser atmosphere in the past.

How do we choose among these various alternatives? We need to find the nitrogen abundance but it would be even more informative to get the ratio of argon 40 to argon 36. The A^{40} comes from the radioactive decay of potassium 40. Argon is very abundant on the earth; it is the third most abundant permanent gas. Ninety-nine percent of it is A^{40}. Now, right off the bat, we can conclude that Mars can't have differentiated and outgassed to the degree the earth has, unless there has been considerable atmospheric loss mechanism, because if Mars had–regardless of chemical reactions–it would have the same amount of A^{40} per unit area as the earth, namely about 5 mb. and there isn't room for that much argon in the spectroscopic observations we have now. This means either that Mars hasn't differentiated and outgassed to the extent the earth has or that some kind of loss of gases of rather high molecular weight must have occurred.

These two measurements, argon 36 and argon 40 will give us a very good indication of the degree of outgassing of the crust, enabling us to narrow down and perhaps even choose conclusively among the various available alternatives. Argon 30 to argon 40 ratios found in meteorites are much more equally balanced.

MURRAY: Toby has pointed out a very important problem. Could an event have occurred in the history of Mars that led to the loss of a terrestrial-type atmosphere? I remind you that there are two cases in getting rid of gases. It is easy to get rid of A^{40} as fast as CO_2 where there is an exosphere that escaped from the atmosphere way above the surface, for example as in the case of the moon. It is much tougher when the exposure is right above the surface. There, weight makes a difference. The reason that A^{40} is a particularly sticky wicket here is that we certainly can't imagine a mechanism that would have selectively removed A^{40} from the exosphere.

MARTIAN ATMOSPHERE FROM MARINER 6 AND 7 OBSERVATIONS

MURRAY: Conway, do you want to continue?

LEOVY: Unfortunately Charlie Barth, who has a great deal to say about some of the key questions, couldn't be here (Barth 1969; Barth et al. 1969).

Barth's experiment, the UV spectrometer, looked primarily in the region from 1,200 to approximately 4,000 angstroms. These observations relate to the problems of atmospheric hazes and the presence or absence of ozone, and the presence of nitrogen or other components in the upper atmosphere. He found spectra corresponding to three constituents in the upper atmosphere: carbon monoxide, CO_2^+, atomic oxygen in the excited ^1S state.

These constituents are all consistent with the simple model of the photochemistry of the Mars atmosphere. The only photochemical model consistent with the observations appears to be one in which only CO_2 is being ionized in the upper atmosphere to a small extent. That is—even though the CO_2 dissociates, it somehow recombines. The mechanism of recombination in the upper atmosphere is not understood but apparently it recombines so that CO_2 in the upper atmosphere is essentially undissociated.

We also have profiles of the electron density from the occultations which can be compared with the CO_2 emission. The CO_2 emission comes from direct excitation in the ionization process, fluorescent scattering, and electron impact excitation. These contributions need to be separated out to get an estimate of the ionization rate (Dalgarno et al. 1970; McConnell and McElroy 1970). The ionization profiles and the UV spectra can both be interpreted in terms of an essentially undissociated CO_2 upper atmosphere.

If we had a model of upper atmospheric photochemical and ionization processes to be confirmed with Mariner data, then we could solidly calculate the temperature of the thermosphere and the base of the exosphere. The model calculations now give a temperature at the base of the exosphere and in the thermosphere in the range of 450° to 500°K.

Another point is that this upper atmosphere is essentially photochemically similar to the upper atmosphere of Venus. Essentially CO_2 is not dissociated but it is ionized. The models of the upper atmosphere of Mars and the upper atmosphere of Venus should be consistent. This temperature is consistent with our knowledge of the temperature of the upper atmosphere of Venus as based on the Mariner 5 data.

HOROWITZ: Conway, I don't understand the basis for the statement that the CO_2 is effectively not dissociated. CO is there.

LEOVY: CO must be only a very tiny fraction as in the lower atmosphere, 10^4 or so of the actual concentration, whereas we know we are getting dissociation radiation for CO_2. The known three-body recombination process is very ineffective and we don't understand why it is not dissociated.

We know the temperature of the Venus atmosphere and it is quite clear that it is not consistent with the dissociated model of carbon dioxide in the upper atmosphere. The carbon dioxide is in the combined form.

At any rate, this tells us that the temperature is low enough that oxygen would not escape, in fact almost nothing besides hydrogen would escape at these temperatures.

The scale height of ionization above the peak enables one to estimate the temperature at that height, on the basis of ionization in a CO_2 atmosphere. This temperature is about 350°K for Mariner 4 and 450-500°K for Mariners 6 and 7.

ORGEL: Can you tell us how model dependent the 450° to 500°K is? Can we expect it to change 50° or 100° up or down as the model changes?

Many of us don't understand these things at all. In your opinion, is 450° to 500°K to be taken as fairly hard or might you come back next year unembarrassed and tell us that it is really 400° or 600°? What new temperature would be embarrasing?

LEOVY: 1000°K.

ORGEL: Are you prepared to put a plus or minus on it?

LEOVY: I think the 450-500°K for Mariner 6 and 7 dayside conditions is accurate to ±30 percent.

MURRAY: Nobody is going to quarrel with 50°.

LEOVY: If the temperature turned out to be 200 degrees different from this, we certainly wouldn't understand the total chemistry of the atmosphere.

ORGEL: Who cares?

MURRAY: No, what he is saying is that the model would be inconsistent with itself.

ORGEL: I don't know whether the other people feel this way, but as a person absolutely unfamiliar with these things, it is obviously important to know whether the temperature depends on the eccentricity of the person who built the model or whether you would all agree that even if all the X is wrong about this feature of the model, it couldn't be off by more than 2,000° or 200°, or whatever.

LEOVY: If this is the correct photochemistry and photoionization, then everybody would get the same answer here.

ORGEL: But that answer is what we don't know.

MURRAY: Your question will be approximate in 1971 when we will have 200 times the total data, taken at different times and different places on the planet. Right now this is an explanation, not a model.

OWEN: I think 1000°K would cover it.

ORGEL: Plus or minus 500°?

MURRAY: It may change. The worst thing is that we have no guarantee that even if this is the temperature now, it was the same last year. Since escape is very proportional to temperature, a few bursts of high temperature can entirely change the picture.

OWEN: You mean the temperature is not known?

LEOVY: It depends on this model.

MURRAY: Conway is being too tactful.

NITROGEN IN THE ATMOSPHERE OF MARS

LEOVY: Let me talk about nitrogen. Barth failed to observe nitrogen in the spectrum. He had expected that if he were to see nitrogen at all it would be in the range between 120 and somewhat over 200 kilometers above the surface. Barth didn't see any at all. He looked at possible emissions in the upper atmosphere, one possibility was fluorescent scattering. Again on the basis of a model of the fluorescent scattering mechanism, Barth says there is less than a certain amount of N_2 at 120 to 200 kilometers; Dalgarno and McElroy (1970) estimated this amount to be 5 percent. It may be only 1 percent. Accepting that, it is still difficult to determine the nitrogen content of the total atmosphere. The problem is that the nitrogen and CO_2 are going to be diffusively separated at the heights of 130 to 200 kilometers, so for any measure of nitrogen in the upper atmosphere–and from there to a measure of the total nitrogen content of the whole atmosphere–we need to know the atmospheric level at which nitrogen and CO_2 would become mixed rather than become diffusively separated. This is unknown.

HOROWITZ: Wouldn't the lighter nitrogen layer out above the CO_2 and diffusion concentrate the nitrogen?

LEOVY: Yes. The total amount in an atmospheric column of nitrogen is very dependent on the height at which the nitrogen and carbon dioxide become uniformly mixed. This is a question of the highest altitude at which the atmosphere is expected to be turbulent. The lower the height at which turbulence ceases, the lower is the abundance of N_2 in the whole atmosphere. If the atmosphere is turbulent up to the height–187 km–of the highest UV spectrum, then the total nitrogen abundance upper limit is close to the 5 percent of Dalgarno and McElroy (1970). The earth's atmosphere becomes nonturbulent or effectively diffusive at a height of between 100 and 120 kilometers, and a pressure of about 10^{-6} mb. We don't know anything about Mars, nor why the earth's atmosphere becomes nonturbulent at that particular pressure.

We believe the turbulence in the earth's upper atmosphere originates tides, large scale dynamics in the lower part of the atmosphere, and keeps the earth's atmosphere stirred up to a height of 120 kilometers. There is every reason to believe—as I will get to later—that the Mars lower atmosphere will be fully as active and hence we might expect a rather high level of diffuse separation. Making reasonable assumptions based on the analogy of the earth as to the height at which separation and mixing join, Barth suggests the order of nitrogen composition of 1 percent.

ORGEL: Could that be off by a factor of 100?

LEOVY: It is terribly soft, but not off by a factor of 100.

ORGEL: A factor of 10?

OWEN: Certainly it could be wrong by a factor of 5.

LEOVY: Whatever the estimated quantity of nitrogen is it is model-dependent in two ways: both on the mechanism for exciting nitrogen in the upper atmosphere and on a model for mixing between the upper and lower atmospheres. At the present time there is little hope for anything hard on the mixing problem, even if we can solve the problem of fluorescent scattering.

ORGEL: It seems to me unfortunate, if this is true, that most biologists are under the impression that the amount of nitrogen on Mars is less than 1 percent, and if the theoreticians are terribly wrong, perhaps as much as 2 percent. This opinion inevitably is held by people who don't know.

MURRAY: Let me say what I think Charlie Barth would say if he were here, because I have been present when he has been pushed on the subject several times. There is better justification for his number than this might appear.

He would say that when flying he has had similar experiences in the earth's atmosphere for a long time and he knows the detectivity limit for nitrogen in the earth's atmosphere by the same technique.

The 1 percent number assumes, as Conway said, the same fluorescent mechanism, the same band structure, and the same mixing ratio.

About the mixing ratio, if anything, he has been overestimating the Mars situation. It is difficult to show why CO_2 should be enriched versus nitrogen. It is much easier to understand processes operating on Mars that would provide for more enrichment of nitrogen over the CO_2, so if anything he has made an error by not including some of the extrapolation there. There has been a lot of thrashing about on the band structure, but so far no one has come up with a case to challenge this.

OWEN: That is incorrect.

MURRAY: I was at that same meeting and was not impressed with what McElroy had to say.

OWEN: The reason was that he was cut off before he could get to the nitrogen discussion.

ORGEL: May we have the discussion now that was cut off then?

LEOVY: But neither McElroy or Barth is here.

OWEN: It is important to say that there are very competent people who feel that this 1 percent number for nitrogen is too small by a factor of at least 4.

ORGEL: ". . . a factor of at least 4"? Are there some fairly competent people–

LEOVY: My personal opinion, just to make matters worse, is that Barth has been conservative in talking about the mixing height. The mixing is a very soft problem, but dynamics happens to be my field. I would be surprised if the actual turbopause isn't higher on Mars, or at a lower pressure, than it is on the earth. As far as the mixing problem, I think it is very unlikely that it would cause the nitrogen to go up, but it might cause it to go down.

MILLER: By how much, if the turbopause is raised a little?

LEOVY: The Mars atmosphere should be more active than the earth's and it should have a higher turbopause, and if you have the mixing ratio–then his upper limits would be reduced by about a factor of 2 or 3.

MURRAY: The mixing ratio is not going to change his limit by a factor of 10.

MILLER: But the weak spot is the photochemistry.

ORGEL: In the photochemistry of N_2^+ is there a strong dependence of fluorescent efficiency on oxygen concentration?

LEOVY: I don't think so.

OWEN: But this is precisely the area in which McElroy–

MURRAY: McElroy also said that he doesn't believe in bands. McElroy doubts the usefulness, in the case of Mars, of the bands. Therefore he argues their absence is unimportant. Other bands he claims ought to be more sensitive. McElroy is throwing stones but it is not clear which way the stones go. Barth might set a much better limit than 1 percent. The photochemistry, difficult to simulate in the laboratory, hasn't been done and the answer is not clear. The limits right now, if we just understood it, might be 10 percent.

OWEN: The point to make here, now, is that it may also be 5 percent.

MURRAY: Right. The same data may eventually show an even stronger upper limit for nitrogen than presently interpreted but still the argument remains constant: We have to prove nitrogen is there.

OWEN: Leslie is asking us what kinds of limits we can set.

MURRAY: Barth said that 1 to 2 percent was his upper limit for N_2

and you want to widen those limits. It may be as low as 0.1 percent or as high as 5 percent. I am just trying to show that there is nitrogen. We would like to find nitrogen so we ask, How can nitrogen be squeezed in instead of looking at this observation that tells us about the probability of nitrogen.

OWEN: But you are biased from the other standpoint.

MURRAY: The real question is what the probability is of one percent N_2 or 5 percent or 0.1 percent.

HOROWITZ: But biologically there isn't any difference between 1 and 5 percent.

OWEN: The point is that right now it is a very soft number.

ORGEL: It seems to be quite clear that no one knows.

MURRAY: The upper limit for N_2 until this experiment was 60 percent.

ORGEL: And now it is 5 percent.

MURRAY: There is a distribution, the most likely value of the upper limit is still 1 to 2 percent. The most extreme upper limit is 5 percent and there is some possibility that the upper limit now is 0.1 percent.

ORÓ: I'll take the 1-2 percent upper limit.

MURRAY: May we now hear the response from the biologists as to the implications of this? Suppose it is 1 percent; so what? I heard one story from Norm.

ORÓ: What did you say, Norm?

HOROWITZ: Firstly, 1 percent is consistent with the idea that Mars has never had large oceans for long periods. This composition is exactly what the atmosphere of the earth would have been if it never had oceans: CO_2 and a little nitrogen. Secondly, in spite of what was said following Charlie Barth's press conference last August about the necessity of nitrogen in the atmosphere for life–I made comments on that myself at the time–it is true that elemental nitrogen in the atmosphere on the earth would not be necessary for life because we have many sources of nitrogen in the oceans. We have nitrates, ammonia, and so on, but on a dry planet like Mars nitrogen must be circulated–and to move it around it must be in the atmosphere. There is no other way to transport nitrogen. If there really is no nitrogen in the Martian atmosphere–by none I mean less than 0.01–I think this is fatal for biology.

ORGEL: Wonderful. Nitrogen will be kept concentrated. Splendid.

MURRAY: It helps those of us who are concerned with observations to hear your views.

ORÓ: I agree with Norm and I have made this point to Toby Owen.

If nitrogen is connected with life, whether in the form of nitrates or something else, eventually N_2 will be produced and there will be a nitrogen cycle. I concur that if the upper limit is 0.1 or 0.01 percent it is tough for life, and the probability is that it could decrease proportionally to the lowering of that number.

SHOEMAKER: How do you decide whether 0.1 percent or 0.01 percent is the cutoff?

MARGULIS: There are two different issues: The direct use of nitrogen by living organisms and the implications for liquid water; the latter is the most important. If low N_2 implies there were never oceans or open water, that is more significant than figuring out the form of utilization of nitrogen. Organisms can use it in many forms: amino acids, ammonia, nitrate, and so forth.

MURRAY: No, let's be sure we are talking about the same thing. The argument is, if there is a nitrogen cycle involving live forms that are presently on the surface of Mars–nitrate-eating bugs, or something–it is reasonable that out of that ecology will come a gas phase. Over time then an equilibrium between a certain partial pressure of N_2 in the atmosphere, high enough to be used tangibly by the organism, will be set up.

MARGULIS: Fixation of atmospheric nitrogen is significant because of its involvement in the cycle. Quantitatively, however, very few organisms utilize N_2 directly.

MILLER: Suppose there is no atmospheric nitrogen but there are 20 grams per square centimeter spread out on the surface. This means an excess of nitrogen for bugs utilizable as nitrate, for example. Then there would be no problem, no need for transport via the atmosphere.

MURRAY: How is nitrogen cycled by bugs? They walk up to a crystal of sodium nitrate–

ORGEL: Microbes reduce nitrate to ammonia and then oxidize it the next winter.

MURRAY: It never goes into the atmosphere?

ORGEL: Exactly, or very rarely.

MURRAY: Then it is a one-way street. But say a molecule of ammonia or nitrogen is released into the atmosphere, what brings it back out to keep the pressure down?

MILLER: If there is much nitrogen metabolism some inevitably will escape as N_2, but at what rate? And what is its rate of reoxidation back to nitrate in the atmosphere? Probably no one here can make a realistic calculation of those relative rates. If the rate of oxidation is much more rapid than the rate of resolution, the steady state concentration may be very low, as low as 0.001 percent or any other number.

MURRAY: We know that the abundance of oxide is very low in the Mars atmosphere. So if these are also going to make the nitrogen very low, it would seem that the rate of oxidation would be very, very low and there would be two very serious contaminants and to get fission there will have to be a product of interaction.

MILLER: First is the rate of oxidation of nitrogen, of NO or NO_2, followed by the second rate constant which is the formation of nitrate. There are no nitrogen oxides over the Chile desert where sodium nitrate is in enormous quantities.

ORGEL: I don't understand the basis for your statement.

MURRAY: Let me give an argument. Somebody should have pointed out that CO_2 is only 0.03 percent in our atmosphere and yet is clearly essential for life on earth. Here we have a small fraction of a critical gas—why can't this be true for N_2 on Mars?

Had someone asked, I would have answered: CO_2 is stabilized by the oceans and if it were not for the oceans the CO_2 would build up—presumably coming closer to the oxygen concentration. The greatest mass of CO_2 is in the oceans where it does the stabilizing.

MILLER: I don't think an ocean is essential to take CO_2 out of the atmosphere. Minerals like calcium oxide will take it up. The ocean is the mechanism here now, but it may not be the only one.

ORGEL: Some of these arguments can be backed by equilibrium constants and rates and so on, and others which we should ignore are backed only by sort of religious intuition. We must only permit discussion backed by arguments. When uncertain to the order of 10^{15} of the rates of the individual processes, don't bother to even bring it up.

MARGULIS: Biologists agree—in the presence of liquid water—that terrestrial organisms could still cycle nitrogen into living things, with atmosphere N_2 cut out. Atmospheric nitrogen per se isn't needed, is it?

MURRAY: But water is needed.

MARGULIS: That is my point. What is the relevance of such a small amount of nitrogen to the water question?

MURRAY: Water is not present now on the surface of Mars so whatever organisms are there now must interact through the gas phase.

ORGEL: Water is another problem entirely.

HOROWITZ: If the concentration of nitrogen in the Martian atmosphere is negligible, if all the nitrogen that has ever outgassed from Mars is in surface minerals, then Martian biology is in trouble. Life there is not necessarily impossible but very special models of biology and geological processes must then be considered.

MILLER: What's wrong with that?

MARGULIS: Don't you mean it is wrong for biology because of the implications for water?

MURRAY: No. Possible systems of life were forced to choose a very narrow spectrum, ones that lead to extremely low partial pressures of nitrogen.

ORGEL: Organisms on Mars are looking at us now and saying, "How incredible they can survive a great thick atmosphere made of 80 percent N_2. They must have evolved wonderful mechanisms to survive."

MARGULIS: To say nothing of surviving oxygen.

LEOVY: The nitrogen question doesn't seem very critical for biology.

HOROWITZ: There has to be some.

MURRAY: Since I don't feel a personal stake in finding life on Mars, I don't really care if all the negative indicators keep coming in negative. But you people now say it really wasn't very important after all, and I just wonder what has been going on.

ORGEL: I think the nitrogen estimate is extremely important, but for a different reason. If something is known about the nitrogen laboratory experiments can be done. It helps specify the soils and soil nitrate concentrations in equilibrium with atmospheric nitrogen and so on. Let's not take the view that life is possible or is not possible. We must see what could go on within the set limits. One extremely important aspect of these results is the suggestion that all nitrogen on Mars may be present as nitrate. If correct, these results have already had an extremely important influence on our thinking about life on Mars. Namely, we have changed our mind—or Stanley Miller and others have changed their minds—from a belief that life on Mars, if it depends on nitrogen, will depend on a combined form of soil nitrogen rather than atmospheric N_2.

As the limits of nitrogen become more and more known people will estimate better the sorts of metabolism consistent with Mars. This changes the sort of experiments done in the life-detection work. This is not extremely important for the statement: Martian life is or is not possible. It is important to exclude possibilities incompatible with the facts.

RICH: We fall into a trap of geocentric thinking—I am just reinforcing Leslie's comments. We really want to know the limits within which we can think about processes which could provide the basis for a living system.

We cannot make a meaningful guess about the probability of Martian organisms with radically different kinds of metabolism and our point here is to see if we can get enough information to modify the questions we ask. If we had a radioactive isotope of nitrogen, comparable to C^{14}, we could do very meaningful experiments. It is a great pity that because of this severe technical limitation we can't really ask the kind of question about nitrogen metabolism we would like to.

MURRAY: Would it be significant if we could rule out the presence of nitrates as a major surface component?

ORGEL: Very significant.

MURRAY: There are terrestrial bands in the infrared. It happens to be a very diagnostic mineral.

MILLER: It depends on the amount; all that is needed is for the organisms to get enough and use it. A tenth of a percent might be quite sufficient, and I doubt if your infrared bands would be that high.

ORGEL: Nitrogen figures are insignificant because they suggest that one now work hard on the infrared spectrum of soil nitrates to try to find limits which would never have been thought of as little as 2 years ago.

MURRAY: I confess to a longer-range view–I want to know what it will take to observationally rule out the possibility of life, or to make its probability very low.

RICH: Short of going to the planet?

MURRAY: You can't do it by going there, you would get negative results. Although some of you may not agree, your colleagues say, "You must go someplace else." It is endless.

KLEIN: No nitrogen in any form would be significant. All the nitrogen on the surface may be in reduced form–basically organisms use it as ammonia or amino acid groups and just turn it around in reduced form. There is no reason at all why organisms must oxidize nitrogen.

HOROWITZ: But if amino acids are exposed to the UV flux of the Martian surface, they decompose and N_2 is produced.

MILLER: It depends on whether the organism is protected from the UV.

HOROWITZ: But when the Martian organism dies what happens to the nitrogenous compounds? Are they protected from UV for 4.5 billion years? You are postulating an enormously tight, closed system.

MURRAY: This confirms my deep-seated suspicion, almost a paranoia, that it is impossible in the next two decades to conduct a meaningful set of experiments to really remove the possibility of life on Mars.

ORGEL: I agree.

MURRAY: And it ought to be recognized: It cannot really be proved that life is not there.

MILLER: Let's talk about proof. If on the first Martian landing there are negative results, look a second, and a third time. Then you begin to estimate probabilities. It never will be reduced to zero but–

MURRAY: This is a religious kind of question, like trying to prove God. You can't disprove life on the moon but nobody is very worried about it.

MILLER: After the third or fifth attempt on the surface of Mars if you don't find life–

MURRAY: Norm is saying this: When the probability of nitrogen is lowered then certain categories of hypothetical life are ruled out. You say this is not relevant, but–

MILLER: Nothing is ruled out.

ORGEL: Not that it isn't relevant, it is not conclusive.

MURRAY: Are you saying there is no such thing as conclusive evidence?

MILLER: What degree of certainty do you want?

ORGEL: This type of issue is not solved by a conclusive experiment. An accumulation of data which gradually persuades people is necessary. If after 10,000 landings you found nothing–

MURRAY: They would then say the wrong experiment was done, the wrong bug was looked for, and if you look for the wrong bug at 10,000 sites it will never be found.

OWEN: Perhaps this discussion should go later with the philosophy.

HOROWITZ: Let me ask a scientific question, no editorializing allowed. Is the stability of nitrates or ammonium compounds in the UV known?

ORÓ: That is the right question.

HOROWITZ: Does anyone know the stability?

MURRAY: Conway, why don't you discuss Barth's data?

STROMINGER: I am a biologist with no feeling for the reliability of the techniques. Can spectral data be taken of the earth on a vehicle traveling to Mars? Can the composition of the earth's atmosphere be accurately measured?

LEOVY: Yes, Barth (1969) describes such experiments, with a rocket not a spacecraft, but with his instrumentation. This can be done for the earth.

MURRAY: That is where the 1 percent N_2 on Mars came from.

LEOVY: Yes, and partly by analogy to the earth's atmosphere. I think Barth's experiments are very good experiments in that comparison with the earth is well understood.

LOWER ATMOSPHERE, HAZE AND BRIGHTENINGS

LEOVY: There are some data from the lower atmosphere of Mars in the UV in the spectral region of 2400 Å. There is very strong evidence that even down at the short-wave length end, 2000 Å, the spectrometer saw the surface. The most conclusive indication of this is the fact that as soon as the spectrometer goes over the edge of the polar cap, there is a very sharp jump in the intensity indicating apparent increase in surface reflectivity.

The opposite end (2000-4000 Å) appears clear. The data, however, do not easily fit to a model which includes a radar scattering component which in the optically thickest cases corresponds to about four times the depth of a radar scattering atmosphere. This means there are observations which can be interpreted as a radio scattering haze, which are variable across the planet, with an optical depth of up to about four times that of a radio scattering atmosphere. Nothing that Barth has seen is interpreted as evidence of ozone absorption in the spectrum. This is very significant relative to his observations and the OAO observations.

MURRAY: Does the radio continue at an increasing rate at the margin of 2,000? Are those consistent between Barth and OAO? We saw a slide from OAO which said the data could be fit with 6 mb atmosphere of CO_2 but you told me they got four times that on the Mariner.

LEOVY: He observed an apparent radio scattering, a variable lambda to the fourth power.

MURRAY: Haze wouldn't have that effect unless the particles were 2 Å in size.

LEOVY: Although there is much interpretation in those data there is good evidence that he sees the surface at least over large portions of the planet as low as 2,000 Å.

MURRAY: The interpretation of Barth's data is unambiguous here: Radiation as short as 2000-1900 Å is reaching the surface. There is no shielding whatsoever down to that level.

LEOVY: Other evidence on the lower atmosphere is from the occultation experiment which looked at the index of refraction data. Under ideal conditions this can be inverted to give the vertical distribution of pressure and, by a second differentiation of the data, a vertical distribution of temperature. Before I discuss that, I want to talk about the TV picture data as related to the surface.

Mars is an extremely interesting object, relative to the moon. Although it resembles the moon, it is variable in the time of day, variable from one opposition to another, and variable from day to day.

One of the most interesting kinds of variable features is the polar cap. The south cap was clearly seen by Mariners 6 and 7 (Sharp et al. 1971;

Leighton et al. 1969).

A classical mystery about Mars is the so-called "blue haze"–the disappearance of surface contrast of features seen in ground-based blue photographs. The Mariner TV pictures, showing topographic contrast in the blue, indicate "blue haze" is not due to any kind of optically thick atmospheric phenomenon, but to the intrinsic lack of contrast in the blue between the dark and grey areas.

It has been convincingly demonstrated from the infrared spectrometer data that the polar cap itself is composed of carbon dioxide. Since no diurnal variations in the edge are seen in the Mariner pictures there must be significant thickness of deposition of frozen CO_2 there, greater than millimeters. Otherwise diurnal variation would be seen.

RICH: The series of four or five concentric arcs going right up and down, are these real or artifact?

LEOVY: Artifact, it's a residual image from the LEM, from the preceding frame.

MARGULIS: If a few millimeters is the lower limit for the size of the polar cap, what is the upper limit for its thickness?

LEOVY: On the basis of models, once how much CO_2 is expected to condense out is calculated over the season as an upper limit to the amount of CO_2 you may get a model dependent number. Some tens of centimeters, I believe.

MURRAY: In the pictures, the topography you can see is of the order of ten to hundredths of meters. That has not been varied, so this is an upper limit. The physical upper limit assumed by radiation exchange is much lower, about a meter.

ORGEL: Could water be in there as well? Is it possible to say?

LEOVY: It is not possible to say.

MURRAY: Yes, it is. There are very strong indications that the cap is pure CO_2 at least from the edge into near the center. The infrared spectrometer was the solid CO_2 spectrum around 2 microns. That is camouflaged and obscured by very small amounts of water, almost a part in a million, and that is, indeed, the reason that the investigators were confused and interpreted it as methane and ammonia. It is a very hard experiment to perform. The fact that it is observed so clearly is prima facie evidence of water's extremely low abundance mixed with CO_2.

LEOVY: Water must be low. In the laboratory spectrum of mixed water and CO_2 the effect of water in part per million is clearly seen.

MURRAY: But it took heroic efforts to get the water out to even see that low amount.

OWEN: How dependent is this on the depth of the CO_2 layer? Suppose water is condensing out first?

LEOVY: Can't there be ice underneath? Water isn't seen, and the stuff acting on the surface scattering radiation back is CO_2 at better than 10^4.

We can get into physical arguments here but the evidence is for pure CO_2.

HOROWITZ: At one of the first press conferences it was said that the laboratory spectra best matching what was seen over the cap consists of about a micron of water ice covered by about 40 microns of frozen CO_2.

MURRAY: Some people don't understand the problem. There are other data for which this experiment is well known; the 3 micron region is one at which water in the system can't be seen.

ORGEL: Is that because water comes on the top of the system?

MURRAY: No. That is the ordinary coefficient, either the gaseous or solid stage. Clearly it is very dry CO_2 on and along about 50 percent of the exposed surface of the cap.

LEOVY: In pictures of the atmosphere, we were looking for discrete clouds, since "clouds" have been observed in ground-based observations. We were also looking for and found a lot of evidence of haze. In the Mariner 7 limb crossing, in the high-resolution frames, we saw something indicative of a haze layer 15-25 kilometers above the limb itself. At 10 times less magnification in the lower-resolution pictures taken in 3 colors–blue, green, and red–we saw the limb double effects and apparent haze, also.

HOROWITZ: It didn't reproduce well in the *Science* (Jan. 1970) paper; they look much better here.

LEOVY: Taken together with far encounter observations, there is a lot of evidence for haze layers. However,the only place where we can unambiguously identify hazes is on the limb, indicating that the hazes are optically thin. The surfaces can be seen very clearly and estimates of optical thickness, within a few hundredths, can be made for the Mariner 7.

In looking for discrete clouds over the surface of the planet in near encounter, we don't see them. The closer one looks at the picture the less one sees. There are some funny looking non-topographic looking bright areas over the polar cap, but nothing clearly indicated to be a cloud.

The most suspicious cloud-looking things are seen in the north polar cap region. Regions in the north polar cap appear to be covered by perhaps more dense haze; there also are some discrete patches that may either be clouds or surface features in the north polar cap region. One classical case is the region of so-called "W cloud" which resembles an inverted W as it approaches the afternoon limb. In ground-based observations a number of areas on the planet are frequently seen from the ground to brighten dramatically as they approach the afternoon limb. Some of these, elements of the "W cloud," were seen as bright patches in the late afternoon (Leighton et al. 1969; Leovy et al. 1971).

MILLER: How do you see a "W" there? [pointing]

LEOVY: It is inverted, the features seen by Mariner form portions of an "M".

HOROWITZ: It is an M in these pictures, but in the telescope it is inverted. That is why it is called W.

LEOVY: My preconceived notion, based on its diurnal brightening, was that this feature behaves like convective clouds. However, there is nothing in any of the pictures in this region that would indicate any structure expected for a convective cloud. The features appear very closely related to the topography and can be traced back over the rotation of the planet. The bright patches can be related to specific characteristic topographic features, specifically to circular features resembling large craters. The "W cloud" appears in the Mars tropics; we also saw bright features a few hundred kilometers in extent, near latitude 40° north. These varied in size between Mariner 6 and 7, but did not show much diurnal change (Leovy et al. 1971).

ORÓ: This is definite geocentric biasing, dubbing them "M's." Who knows what the Martian alphabet is?

LEOVY: Another area, observed by ground-based work to have variable brightening, is the feature Nix Olympia (Leighton et al. 1969). This feature appears to be a crater with multiple concentric rings. Multiconcentric quasicircular areas are seen in other nearby areas. The bright rings don't appear to be convective clouds. They brighten later in the afternoon than is expected for convective clouds and they display no structure expected in connective clouds. They seem closely tied to the topography suggesting they are either on, or very close to, the surface rather than well up in the atmosphere. The convective area on Mars ought to be about 15 to perhaps 25 kilometers deep by early or mid-afternoon of the Mars day. It would be surprising to have anything at that height in the atmosphere closely tied to the topography even as seen at relatively low resolution.

What is happening? Is this a condensation mechanism? The time of day and the latitude precludes the phenomenon being CO_2 condensation unless it is high above the surface. Twenty five kilometers is about the lower limit to the height above the surface that it would have to be if it were CO_2 condensation.

From temperature profiles deduced from the occultation data at the entry and the exit of both spacecraft, we can try to see whether what we are observing in the atmosphere is consistent with CO_2 condensation. These data indicate that the temperatures are warmer than is required for CO_2

condensation, up to 25 kilometers (Rasool et al. 1970). Mariner 6 occultation data were subject to some drifts and the temperature results, since they come from a second differentiation of refractivity data, are somewhat questionable.

The infrared spectrometer observed the reflection spectrum of solid CO_2 in the atmosphere of Mars above the limb. Two of these limb crossings were in the tropics. The reflection spectrum of solid CO_2 was observed over the limb in all crossings even though the places where the solid CO_2 appeared in the spectra corresponded to a different height above the limb in each crossing. In Mariner 7, as reported by Herr and Pimental (1969), the second limb crossing height of the spectrum of solid CO_2 happened to occur at 25 ± 7 kilometers.

What can be said theoretically about the temperature and the structure of the Mars atmosphere? From the radiative properties of CO_2, the solar radiation, the surface temperature, and the extrapolations from terrestrial experience we can estimate the lower atmosphere temperature. On this basis, as well as from the radio occultation data, we infer that the bright patches of the "W cloud" are unlikely to be solid CO_2.

ORÓ: What is it?

LEOVY: I don't know.

MURRAY: Conway has now told you—for better or for worse, it is very exciting.

ORÓ: What is the peak?

MURRAY: It seems to be on the ground. They are perfectly reproduced from the Mariner '69, as far as the occultation data.

OWEN: What is your conclusion about the brightenings on the ground?

LEOVY: I don't think they are a CO_2 cloud, and I think they are close to the surface.

MURRAY: He thinks they are on the ground but we cannot understand how they could be stable on the ground. From the pictures they appear as actual surface deposits.

OWEN: They don't have to *be* on the ground, they could be ground fog.

LEOVY: They could be close to the ground. As far as the physics of it, there isn't very much difference.

MURRAY: They are equally implausible.

SÖDERBLOM: Why? If the topography of Mars is so strongly controlled by the dynamics of the atmosphere, particularly if objects are about 1.5 scale heights, a localized transient phenomenon could be keyed to the topography.

LEOVY: If these are of the order of 1.5 scale heights, you are right, but there is no real evidence of scale heights that size although they may be there.

RICH: Aren't these features appearing at this resolution in the same area where we have higher resolution photographs?

LEOVY: Unfortunately we don't have high resolution photographs of this region.

RICH: I know, but aren't there similar features in the region where we do have high resolution?

MURRAY: Because of the mysteries of the relation between Mars' spinning rotation time and that of the earth, there was a bias on the selection of the sites for photography. Unfortunately there is no coverage of the entire area.

OWEN: In answer to Larry's point, there is a ground-based observation of a region called Albor, in Elysium, where brightening is observed periodically; this is in the radar swath and is definitely high.

SODERBLOM: When looking with the radar at the topographical irregularities there is a resolution of 300 kilometers, but we are talking about scale heights that are different to the order of 15 kilometers, so we are talking about the low frequency nature of the topography. The high frequency nature could have huge variations.

LEOVY: Even if that is the case, then the suggestion stands that it is related to a local source–condensing out in that area rather than a mountain there causing enhancement of convective activity. It seems unlikely to be carbon dioxide concentration unless it is above 25 to 30 kilometers high. If it is water condensation, there is a problem because with 15 to 30 microns of water, the water, in the long term average, can be in equilibrium with the average ground surface temperature at latitudes above about 45° or so forward on either side.

At more tropical latitudes the water vapor is going to be evolved comparatively rapidly from the surface. Its layer of permafrost is below the surface. Within a period of a few hundred years it will be vaporized to such an extent that further evolutions would be reduced essentially to zero.

The only way to maintain a pure water model of these reactions must be a water flux in conjunction with a local heat source. The heat must continually force the water up, assuming a very large supply of frost. This would resemble somewhat the geotherm of the earth.

Even in the case of the geotherm area, the upper limits to the rate at which water can evolve cannot be calculated because any water flux tends to be pinched off by freezing in the soil. Given a heat flux, a proportional upper limit to what the water vapor flux can be—due to the effect of the pinching off—can be figured out. If you assume heat fluxes that are related to the earth's mean heat fluxes as standard, the kind of upper limit water evolution in a region like that is about a few microns precipitable water equivalence per day. Let me stop my talk at this point and see if there is any discussion.

ORÓ: Weren't you going to tell us about the haze?

LEOVY: The limb haze, seen in the Mariner 7, could not be carbon dioxide if the occultation data on temperature are correct. If the data are incorrect, it could be carbon dioxide ice. This would then be consistent with the infrared spectrometer observations of frozen carbon dioxide over the limb.

It is just plausible that the surface heated convectively by radiation could be sufficiently cooled at a height of 25 kilometers in the atmosphere to get a CO_2 haze at that level. The hazes we see therefore could be carbon dioxide, but they could be water or even dust layers in the atmosphere. At the moment we can't really rule out any of these possibilities.

HOROWITZ: Pimental said he saw some bands of silicate above the limb.

LEOVY: That is right.

MURRAY: Unfortunately, these observations have never been published or displayed, so we don't know. We have depended on oral descriptions of data which have differed at different times. It is hopeless. They either are or are not silicate bands; they either are or are not CO_2 bands; or perhaps they are both.

OWEN: It depends on the interpretation.

MURRAY: Unfortunately, I think it is seasonal. I think a transitory phenomenon is involved. I have also heard about strong silicate absorption bands in the data on the planet itself, which if true would be very interesting. Since these data have not been made available we don't know what is there.

LEOVY: In summary, if water is one of the more biologically important issues, then these areas of tropical transitory brightening are particularly interesting and call for closer investigation. The variably bright features around the north polar cap where we have never been able to get very close could either be CO_2 or ice. But the discrete bright features in the tropical areas are very unlikely to be CO_2.

OWEN: Is one reason that no clouds were seen because of a dependence on Mars' position in its orbit that has been established and when it is closer to the sun there are more clouds or something?

LEOVY: Cloud sightings are so few, and observations of Mars over its orbit are so badly distributed that the statistics of cloud occurrence are not very reliable.

OWEN: Were you perhaps unlucky in that encounter because you didn't see the clouds?

LEOVY: Ground-based observations classically see two kinds of cloud observations, the so-called white areas, most stationary or nearly stationary, such as the so-called W cloud, which may not be a cloud at all but a surface feature.

Then there are features that definitely move, most of these are the so-called yellow clouds. They are very rare, even in ground-based work. They are so rare that we would have been incredibly lucky to have seen one, but not so rare that they are of no significance. In the 1956, and I believe the 1926 Mars opposition, the planet was essentially covered by these yellow obscuring things, whatever they are. The atmosphere is very active with respect to the surface and to get anything, no matter what, that appears optically thick in the atmosphere must be very significant on the surface.

MURRAY: A third category that may contribute to the confusion is in blue light. There are bright patches seen frequently. Some were seen in the general vicinity at the time of the fly-by. Some of these blue things in the later far-encounter data of this spacecraft were seen, too. They may be thin hazes, but in the blue they become brightly scattering enough that they really begin to be two different things. They are not clouds in the sense we think of water clouds or something optically thick.

OWEN: Hasn't Baum studied a lot of moving white clouds?

LEOVY: Most—about 90 percent—of the clouds he looks at are stationary and in particular most of the white clouds are stationary. Because they didn't specifically try to differentiate between white and other colors of clouds, interpretation is a bit difficult.

Mars, the time we saw it, was rather inactive in ground-based observations. There were some indications of brightening but the W cloud, for example, was much less bright than it very often is.

HOROWITZ: Was Hellas bright from the earth at the time Mariner was observing it?

MURRAY: No, it was not bright.

The Mariner data have given us some greater perspective on these troublesome brightenings. They are not high in the atmosphere if they are in the atmosphere at all, and they have a very great dynamic range. Remember Toby's statement about this feature in Elysium that got brighter on a daily basis as the subsolar point moved up and this funny phenomenon developed? Ice might be producing water. To me, this is one of the best clues for the existence of permafrost.

We found more, these brightenings are definitely there and probably are not ammonium chloride, which is a nice bright color. Our choice narrows to CO_2, ice or water by a process of elimination. The CO_2 won't stay frozen at the high temperatures in these tropical areas. We are almost forced into thinking water vapor is involved, particularly in those brightenings that occur at the time of the highest temperatures in the tropical areas. Frozen CO_2 is probably responsible for brightening at the lowest temperatures. Is this a fair summary?

LEOVY: The temperatures from the occultation agree very well with the theoretical ones. If the temperature distribution is correct, then the dynamics of the wind systems of the Mars' atmosphere can be worked out in detail. One expects on this basis that the lower atmosphere of Mars is very active dynamically and has rapid mixing rates as compared with the terrestrial atmosphere (Leovy and Mintz 1969; Leovy 1969; Ingersoll and Leovy 1971).

MURRAY: To conclude this morning's session I make the following comments. We haven't discussed the polar cap question, which deserves attention. I propose we take that up right after lunch.

Nobody has said anything about the carbon suboxide hypothesis.

HOROWITZ: We also haven't covered the light and dark areas yet.

MURRAY: Yes, but with regard to the stability question of the volatiles we need to log in our awareness of the carbon suboxide hypothesis. Stanley, in particular, has given this a good deal of thought. Perhaps we can discuss it later, too.

Looking at our list, nothing relevant to the debris mantle has been said this morning. With respect to the water cycle, Toby has told us about amounts, and Con's discussion told of these brightenings which certainly point a finger at a possible role of surface water.

OWEN: I must remind you that this very coarse number, 15-30 microns, if not a planetwide average is at least a hemispheric average. It should be regarded only as that. Obviously, local variations may be considerable with a substance like water.

MURRAY: Of course. But if the variation was very large, water would have been seen in the on-board spectrometer on Mariner 7.

LEOVY: Also, Barth observed lambda and alpha in the exosphere of Mars, and 2 weeks ago at the meeting in New York he presented an interpretation of lambda alpha in terms of an escape rate for hydrogen of about 10^8 H per square centimeter per second. This number is in reasonable accord with these amounts of water, if UV gets down to the surface and dissociates.

MURRAY: Did he give a mean lifetime of the water molecule on the surface?

LEOVY: The mean lifetime for a water molecule is about 10^7 seconds.

MILLER: How many grams per square centimeter would be lost?

LEOVY: That would get rid of a few meters, about 10^{25} molecules of water—that is, the liquid equivalent of water—in a 4.5 billion years.

OWEN: If at an earlier era there were less CO_2, the shielding would be less effective so that rate would go up.

LEOVY: It is not very strongly shielded. Water is a stronger absorber in that part of the UV anyway.

OWEN: But the maximum absorption of the water is 1850 Å and the CO_2 at the surface is shielding to 2,000—

SHOEMAKER: How good is this number?

LEOVY: As I remember Barth's statement, it is better than an order of magnitude accurate.

MURRAY: If we could ever understand the Venus escape rates and what the chances of outgassing are.

OWEN: It is still a continuing problem on the earth.

MURRAY: Let's go on to the nitrogen cycle.

HOROWITZ: But it is time for lunch.

SHELESNYAK: We ought to break right now.

The conference recessed at 12:25 p.m.

Sunday Afternoon Session

THE MARTIAN SURFACE

MURRAY: A criticism of this morning's session is that we have been too polite and quiet, so, short of physical combat, do express yourself as fully as your oral talents permit.

SHELESNYAK: However, the communications system should be limited roughly to one conversation at a time. When we get three or four we just get noise.

MURRAY: O.K.

We have been talking about the atmosphere. Next we ought to briefly consider the solid portion of the planet in equilibrium with it. We should discuss the solid-vapor equilibrium, particularly the polar cap and water, the carbon suboxide hypothesis and from there go into the permanent surface features and their implications.

SCHOPF: Have you decided not to go down the list?

MURRAY: You're right, going down the list provides us with a way of regurgitating what we discussed before.

The nitrogen cycle is based on negative results. What does the absence of N_2 imply? It rules out cycles that imply about 10 percent or greater partial pressure of the Martian atmosphere. At the moment this is all we can say. It also rules out equilibrium concentrations as the result of 4 billion years activity that have led to more than 5 to 10 percent composition of the present Martian atmosphere.

It is interesting that in this group concerned with life, there has been very little discussion of the carbon cycle. The carbon dioxide pressure is that of a carbon dioxide atmosphere; carbon monoxide is present only in very small quantities. This probably results from equilibrium due to hard radiation incident upon the surface of the whole planet.

OWEN: If this is an earth analog outgassing situation, then we also have a demonstrated failure of the Urey equilibrium between the carbonate and silicate rocks in the absence of water. CO_2 is all atmospheric and not in the carbonate rock.

As I recall Rubey's discussion, he was hesitant about accepting Urey's chemistry. He felt the situation was more complicated, particularly since a lot of the CO_2 is in limestone, in the form of shells.

MILLER: Yes, Rubey is right, it is very complicated. For those who aren't familiar with this, the Urey equilibrium partial pressure of CO_2 is what keeps the approximately 4×10^{-4} atmosphere. Although more complicated it seems to be in overall control by the Urey equilibrium, except that he wrote down the wrong mineral, but it doesn't change the situation.

The effect of the organisms depositing carbonate is presumably not that big because the calcium carbonate is more or less saturated in the ocean.

OWEN: Isn't it also dependent on erosional processes?

MILLER: Calcium must be supplied to precipitate the calcium carbonate.

OWEN: Fresh surfaces must be continuously exposed.

MILLER: Yes.

MURRAY: Calcium carbonate deposition also depends on organisms laying it down.

MILLER: No. If the organisms do not deposit it there might be twice as much atmospheric CO_2. The calcium carbonate is essentially saturated or a little below saturation in the ocean.

MURRAY: But I think the limestones are the result of organic processes.

MILLER: Yes, most are.

SCHOPF: What happened in the Precambrian before there were invertebrates with hard shells?

MILLER: If the organisms that precipitate it were absent it might take supersaturation by a factor of 2 to precipitate nonbiologically, but that doesn't really affect anything. With respect to the Precambrian presumably the stromatolites, carbonate stored away in algal colonies, caused precipitation. Hard shell organisms aren't necessary.

SCHOPF: Right, but most limestones are not necessarily of biological origin. There are many examples of inorganic precipitation.

MILLER: Which? I am told that the limestones off the Bahamas, for example, have been more or less biologically precipitated. That example is always given.

SCHOPF: Yes, I have that impression, too, but I may be wrong.

MILLER: But with reference to Mars, this is not relevant. If there were no biological precipitation it could occur nonbiologically. Just add some calcium chloride to sodium carbonate and see what happens.

SHOEMAKER: There are examples of inorganic, nonbiological limestone in saline deposits, as well as in certain fresh water but not ocean systems.

MILLER: Are they extensive?

SHOEMAKER: No. Your generalization is good for Phanerozoic stones. Most apparently are precipitated biologically.

MILLER: In any case, it is clear that if the organisms don't precipitate calcium carbonate, it will precipitate itself.

MURRAY: The fact that the carbon dioxide appears in the atmosphere, rather than in the rocks on Mars, suggests that there never were oceans in which this equilibrium took place or that something has happened since the oceans existed to completely eradicate the earlier situation.

OWEN: But we don't know how much CO_2 is in rock. We see some in the atmosphere and no liquid water at the moment. But we could be very wrong in concluding that we are seeing all the CO_2.

SHOEMAKER: What would be the atmospheric equivalent in terms of carbonate if it was all precipitated out?

OWEN: If the earth's limestones and so forth were put back into the gas phase you'd get 30-40 atmospheres.

SHOEMAKER: But we are talking only about a layer of carbonate that is a centimeter or a few centimeters thick.

HOROWITZ: Yes, about 20 grams of CO_2 per square centimeter.

MILLER: So, it is 100 grams per angstrom.

ORÓ: Is there so much CO_2 in the atmosphere because, in the presence of this small amount of water, the available calcium ions have already formed carbonates? Have you actually much carbonate on the surface and is this why there is so much CO_2? Are all the ions CO_2 could combine with already combined under the conditions of the Mars environment?

OWEN: Following the earth analogy, the limit on the CO_2 would be set by the fraction of the atmosphere that is not CO_2. A high proportion was expected to be something inert, not affected by the chemistry, like the argon and nitrogen, Unfortunately, our present uncertainty in the surface pressure is such that we can't determine it.

ORÓ: What is the total percentage of the other gases, not CO_2?

OWEN: We don't know the pressure well enough.

MURRAY: Not even including the data from the occultation measurements?

OWEN: The occultation measurements depend on the spectroscopy for the composition; they tell us 80 or 90 percent of the atmosphere has a molecular weight of about 44.

MURRAY: But argon and CO_2 aren't that different in weight.

OWEN: That is the problem.

MURRAY: But the relative composition of argon and CO_2 doesn't make any difference to a first approximation for the solution of the retardivity problem of the occultation. The total pressure is not very composition-dependent.

OWEN: Yes, but the composition can't be deduced from the pressure measurement. That is the difficulty.

MURRAY: But you say there is a measurement of 6 millibars of CO_2.

OWEN: Because of the great uncertainty in the spectroscopic measurement we don't know if that 6 mb. of CO_2 is at the same level that the occultation is measuring.

SODERBLOM: We really have no idea until we know something about the partial pressure-equilibrium conditions and the carbonates. Either plus or minus in 4 billion years it goes to saturation or zero, so we really don't know.

MURRAY: Can't we use an indirect way? The polar temperatures are reasonably close to those predicted, about 148-149° and that suggests that CO_2 controls it. This suggests 4 to 5 or 6 mb. of CO_2 by just equilibrium of the polar cap. Therefore the atmosphere certainly is not going to be composed of 50 percent something else, implying for example a very small amount of argon.

OWEN: No, if there is 10 percent argon that is a very large amount.

MURRAY: Here we go again. What precision is needed to make the damned measurement meaningful?

OWEN: If we follow the earth analogy, 10 percent argon in the Mars atmosphere is very high. Argon should be a fraction of a percent, which it may well be. The difficulty is that the atmosphere may be 99.9 percent CO_2. Unfortunately, we don't really know.

MURRAY: Let's continue through the list. This afternoon we'll discuss the light and dark areas. Nothing was said this morning relevant to the internal history.

RICH: Incidentally, will the '71 Orbiter detect Martian mass-cons?

MURRAY: Yes, if they are about as large as lunar mass-cons. There are huge gravity anomalies. The problem is that Mars could have the same surface structure as the moon and yet the mass-con anomalies could be smaller than on the moon.

RICH: If the Martian mass-cons were comparable to what is on terrestrial continents, would they be detectable?

MURRAY: Gene Shoemaker may correct me, but I think not. They contribute both in temporally variable and zonal terms. I don't think they would be detected from orbits that high. The minimum, about 1,600 kilometers, may go down to about 100; but mostly the spacecraft spins very far away and these high-order terms tend to die out.

SHOEMAKER: Perhaps I don't understand you, Alex. There isn't a significant free air anomaly associated with the continents, as the so-called mass-con concept suggests. It is a very bad word. What is seen is a mass load supported by the surface of the moon.

RICH: Don't mass-cons produce an effect comparable to what the continents produce in the earth—but larger in the case of the moon?

SHOEMAKER: No, these are entirely different. The continents are actually supported buoyantly. The *mass-con* is a load not supported buoyantly; that is why they are seen.

MURRAY: This is like the Hawaiian Islands, a mass of material that stands up, supported by the strength of the crust. The anomaly to first approximation is as if you took a knife, lifted that piece of mass off, and asked what the contribution was. There has been gravity differentiation of continents, which results just because the residual anomalies are very low.

SHOEMAKER: The continents aren't seen as perturbation, but there are positive gravity-free air anomalies on the earth of the same order.

RICH: What are they due to?

MURRAY: Generally they are associated with large young volcanic fields, piles of volcanic materials supported over a short-time period by the strength of the crust.

The interesting question is if we will be able to resolve the question of the correlation between topographic variation and height variation and gravity. To some extent the answer is yes. However, I still doubt if we approach the precision we have in the earth, and if the planet is well compensated, that we might know it because we don't detect anomalies.

SHOEMAKER: What will be the periaxis difference?

MURRAY: Sixteen hundred kilometers. It is bad.

SHOEMAKER: Any anomaly smaller than that lateral dimension will not be seen.

MURRAY: The apoaxis is 25,000 kilometers.

SHOEMAKER: But you would only see the anomaly; you would still see the accelerations if there were an object of that dimension.

RICH: Isn't the orbit controlled from ground?

MURRAY: Yes. It is injected very carefully and trimmed up, and so forth.

RICH: At the end of the mission would there be a good reason to make a smaller orbit in order to detect these?

MURRAY: The orbit case is already over specified. Many different experiments are on board; about 15 different ways are available to spend the money. There is quite a battle going on. The resolution of pictures, the limit of how low the spaceship will be permitted to go, depends on the sterilization question. It is a true statement that the exploration of Mars will be limited because of sterilization.

SHOEMAKER: What is the lifetime in orbit once it is dead?

MURRAY: Twenty-seven years, by definition, which is part of the sterilization question, with a probability of 10^{-4} and a probability of .9999 that it will be in orbit 27 years.

SHOEMAKER: With a periaxis of 69 kilometers, why should it decay that fast?

MURRAY: Gene, you are awakening wounds. My stomach is flipping over.

HOROWITZ: In the worst case of an atmosphere I guess it would decay in 50 years.

MURRAY: For the second time there is an extreme view of this matter, entirely one-sided, which actually hurts us. The first case was when we did not try the probe. Sterilization was the principal reason it was not done.

Continuing with the list, we have surface radiation. Solar radiation penetrates to at least 2,000-1,900 Å and reaches the surface without significant growth. *Significant* means that 10, 20, or 30 percent of the incident radiation reached the surface because the albedo effect across the pole that is seen very easily in the scan is very high.

MARGULIS: This surface UV provides an ideal natural sterilization mechanism.

MURRAY: This has been noted. The radiation would sterilize terrestrial organisms, but they argue that Martian organisms love it; apparently they gobble it up, growing fat and healthy.

MARGULIS: Then since their evolution has been so different, they couldn't possibly continue to grow on anything terrestrial.

MURRAY: That is right.

MILLER: There are many easy mechanisms for organisms to protect themselves against this UV radiation.

MURRAY: But some organisms must like it. How do they extract needed energy from the solar radiation?

MILLER: This depends on how they evolved. Assuming they have DNA, they can easily protect themselves from absorption and damage at 2600 Å. Protection against ionizing radiation, protons and so forth, is more difficult to see. There are DNA repair mechanisms—with which Leslie Orgel is probably more familiar than I am. Since there are well known ways to repair UV damaged DNA, I don't think it is valid to argue that this radiation will kill organisms that presumably evolved in its presence.

LEOVY: We can compare the penetration of solar protons in the polar region in the earth's magnetic field. Very little charged-particle penetration is below 60 kv telemeters, which is a much lower pressure than Mars. I doubt if it would be a more severe problem in the case of Mars—even without a magnetic field—than in the case of the earth.

MILLER: Then if only protection against UV is needed and this is no problem, the UV argument cannot be used against the presence of life.

MURRAY: But if we require that the organisms must adapt to high UV to low partial pressure nitrogen, the absence of liquid water, for example, the combined adaptations might have limited populations that evolved there.

MILLER: You don't necessarily multiply these kinds of improbabilities.

HOROWITZ: Bruce, that is not a good argument; drop it.

ORGEL: Let's turn it around to see how impossible it must be for life to survive terrestrial conditions. There is no untraviolet life, enormous quantities of toxic oxygen, masses of nitrogen around, enough water to drown out everything. Evolutionary adaptation to those things is clearly not possible.

HOROWITZ: The UV becomes important in protecting Mars against contamination by terrestrial microorganisms. Under the Martian UV flux, no terrestrial microbe will survive on the surface for more than a few minutes.

SHOEMAKER: Can that argument be used?

HOROWITZ: We have tried.

YOUNG: But protect them with a few grains of limonite on top.

HOROWITZ: Yes, that argument is used. The terrestrial invaders will be safe if they never come out in the daytime, and always move around at night.

MURRAY: Their needs to complete their life cycle will be provided by an unestablished mechanism.

YOUNG: They will eat the carbohydrates that Hubbard et al. (1971) found to be synthesized in a Martian atmosphere, working their way toward the surface, but never quite reaching it.

MURRAY: Since you will respond with nothing but religious irrationality on this subject, let me suggest an alternative culture—a test tube culture zapped with hard ultraviolet irradiation.

RICH: A culture of Martian bugs?

MURRAY: Yes. Say we have 11 bacteria growing on Martian soil that have evolved shielding mechanisms. They take what they want and what they don't want is sucked away. They must have some waste products, a chemical ecology of some kind. Can they shield all their by-products from UV irradiation, too? These chemicals must get broken down very quickly.

ORGEL: That is a tremendous advantage, lucky creatures.

OWEN: Their environment isn't polluted.

SHELESNYAK: They can be designed like small halibut, dark on one side and light on the other,

MURRAY: What happens when these long chain molecules are broken down to methane, ammonia, carbon dioxide, carbon monoxide, water?

YOUNG: They reach equilibrium.

MURRAY: Unless the chemistry is very simple some products are not carbon dioxide in water. What happens to them; do they go up in the atmosphere?

MILLER: In one case they would go into the atmosphere and this is what George Pimentel probably was thinking about at his press conference. They may just have a bag that contains the stuff for reuse. Certain deep-sea fish store oxygen in their swim bladders. Some seaweeds store carbon monoxide in a bulb. Storage bags are easy to envision and there are many analogies with terrestrial organisms.

MARGULIS: Some blue green algae apparently have gas vacuoles that are used to regulate distance from the surface.

MURRAY: Again we are invoking another probability.

HOROWITZ: Life is improbable to begin with.

MURRAY: After having had this profound conversation here, continuing with the list, we will get on to the sublime: transitory features. Under inherent transitory features we mentioned the brightenings. Whatever they are, they qualify as transitory features. So do clouds.

We had quite an extensive discussion of anomalous frost and clouds. Summarizing what you said, Con, we didn't find any white patches that we can say are clouds or frost patches for sure, other than brightenings—whatever they are.

The oxygen question was not discussed directly but we said there is no believable measurement of oxygen, only a very questionable upper limit. The absence of ozone in the Mars spectra also placed a pretty good limit of oxygen.

OWEN: Barth's sensitivity on the ozone measurement is still in question, the problem of the ozone-oxygen relationship is comparable to the $CO-CO_2^+$, namely extremely model-dependent. There really is not a unique answer to the question: Given so much oxygen, how much ozone would you have or vice versa?

Because of the small amount of oxygen, if ozone is present, it will be near the surface of the planet rather than 'way up in the atmosphere. That indicates to me that there isn't very much, because ozone is so damned reactive.

MURRAY: What about hydrogen resulting from dissociation of water and boiling off? By the same argument oxygen is produced at the same rate —one oxygen, which presumably doesn't escape, for every two hydrogens. This implies that a small but real amount of free oxygen is generated. How much is it, several meters?

LEOVY: Several meters of liquid water equivalent over 4 billion years.

MURRAY: If it is about a meter per billion years, that is, a millimeter per million years of oxygen is made by dissociation of H_2O alone.

MARGULIS: What is happening to it?

MURRAY: Yes, where is it?

MILLER: In the absence of life on the earth, the oxidation of carbon to CO_2, sulfur to sulfide to sulfate—of which there is a great deal in the ocean—and ammonia to nitrogen would occur starting with reducing conditions. In the case of Mars the nitrogen might be expected to oxidize to nitrate and, finally, the ferrous to ferrite.

MURRAY: How much carbonaceous chondrite has come to the surface during the same billion years? Gene, do you have a feeling for that?

SHOEMAKER: Very little.

MURRAY: We once estimated how much water was brought in to the moon by meteorites. Hypothetically, the carbon in carbonaceous chondrites and comets might have provided enough reduced material to use up the oxygen supply.

SODERBLOM: The pressure of oxide on the surface is very small by comparison with the local oxygen in the atmosphere.

MURRAY: As Stan said, the ferrous-ferrite is an easy reaction to cause that.

SODERBLOM: But extending over 4 billion years, it happens.

MILLER: It is a question of kinetics. The iron reaction usually requires water, but it might go in the absence of water.

LEOVY: If the Mars atmosphere were put on top of the lunar debris layer, with the model layer of stirring, and so forth, would much oxygen be produced?

SHOEMAKER: No, if I am not mistaken in my mental calculation here, I think it comes out to a few milligrams of carbon per square centimeter.

MILLER: But how much troilite per square centimeter? And iron, the ferrous-ferrite?

SHOEMAKER: I thought we were talking about carbon. The troilite would be quite a bit more. There is easily enough iron to be oxidized to absorb the meter of oxygen.

MARGULIS: Have the moon rocks been oxidizing since they came to earth, has any effect of the terrestrial oxygen on moon rocks been seen?

SHOEMAKER: No perceptible effect has been seen at room temperatures. The critical experiments haven't been done. The lunar rocks are quite stable but almost certainly the small amount of iron would oxidize if it were exposed.

MURRAY: Our conclusions seem to be two: (1) there is no known oxygen; (2) the amount of oxygen from dissociation of water is higher than we thought in the sense that the occultation rates are higher. Breaking down of water is higher than expected but the amount of oxygen produced is so low that is represents no problem; it easily can be absorbed by many mechanisms. This implies that water and oxygen cycles do not really have to be related, at least under present conditions.

KAPLAN: Bruce, how much oxygen are you predicting would have been made in a billion years?

MURRAY: A meter per billion years is consistent with the amount expected from the breaking down of existing water into Martian atmosphere. A rough calculation of *how much* can be made from the ultraviolet and the water. It is a small effect that will not change the surface chemistry a hell of a lot except it might affect the color.

KAPLAN: We don't know. The water content of carbonaceous chondrites is very high, at least according to what we measure here, but maybe this isn't true.

SHOEMAKER: The anhydrous minerals are certainly indigenous to the meteorites.

KAPLAN: We measure as much as 20 percent water in some chondrites, but maybe the true content is only 2 percent.

MILLER: Actually, it is lower. The epsomite–magnesium sulfate–probably entered the earth's atmosphere unhydrated and became hydrated on the museum shelf.

KAPLAN: But we don't know how much water is necessarily indigenous.

SHOEMAKER: Obviously they are hydrosilicates.

MILLER: We found two fractions of water with the different deuteriums––

OWEN: If there are many cometary impacts they really just bring in ice–don't worry about minerals.

SHOEMAKER: The question isn't where the water is coming from. It is just that presumably absorbed by the Martian surface, oxygen will accumulate from water dissociation.

MURRAY: Let's continue our review. No data was presented this morning related to surface composition.

Now surface pressure: An implication of the data is that it varies greatly across the planet. One parameter you have to work with is the considerable variation in surface pressure just because of the height variation.

At least for short periods, surface pressure may have changed in the past. Cataclysmic events changing the radiation balance at the polar caps may have driven more CO_2 into the atmosphere. Although farfetched, it is not impossible. From the pictures we see that the surface of Mars has experienced processes, apparently not presently manifested, that may well have been cataclysmic. Since there are no oceans to buffer the surface processes on Mars, things may get way out of balance for a while—such as during dust storms. Mars' history may be like desert erosion: Maybe the present picture is caused by a few very significant events that have been widely separated in time. If so, looking at a single instant in time may not be representative.

It is important to find and characterize different environments that are more and less interesting to the problem of life on Mars.

KAPLAN: Would we expect an acid pH—if we can use those terms—in the surface soil or regolith of Mars? Since apparently there is no liquid water to act as buffer for the bicarbonate-CO_2 system, but there is CO_2, if the soil is alkaline with traces of moisture, shouldn't CO_2 go straight into carbonate form? Plagioclases would weather down and form calcium hydroxides. If the soil were alkaline the carbonate would go right back into it and no vapor phase at all would be expected. Has this been looked at?

MILLER: If there is any water liquid or even vapor the pH will be more or less acid, the pH of a bicarbonate solution. The pH of something really dry can't be measured.

KAPLAN: Yes, I understand. But to understand the surface properties, would we expect neutrality, a bicarbonate solution, or would the pH be on the acid side?

MILLER: If a sample of rock is treated with water in equilibrium with the atmosphere, the solution ought to have the pH of carbonic acid, about 4 or 5. In the atmosphere I would still expect a pH of more or less than 2, although it is hard to judge.

KAPLAN: I agree. I just wondered if anybody has made any theoretical calculations or holds any views on this.

HOROWITZ: If carbon suboxide polymer is precipitating out—as has recently been proposed—it hydrolyzes to malonic acid. It is very hygroscopic. If that model is correct, the surface of Mars must be extremely acidic.

MILLER: If there is a lot of malonic acid, the pH ought to be around 3 or 4, which isn't all that low.

MARGULIS: It is pretty low for origin-of-life experiments. Bruce Murray can use it as fuel for his fire.

HOROWITZ: Is pH 3 or 4 right? It has two carboxyl groups.

MILLER: That is a little more acidic than acetic acid, which would give you—

HOROWITZ: pH 4 to 5.

MILLER: It depends on the buffering.

MURRAY: May we go on to the discussion about carbon suboxide? Perhaps I should summarize the recent results on the polar cap and the CO_2 question and then Stanley can talk about the buffering problem.

MILLER: I am not really expert on the carbon suboxide proposal.

MURRAY: It is an exceedingly interesting idea and if by any chance it should work–it would solve some mysteries and pose some interesting questions. It ought to be looked at rather hopefully anyway.

We discussed considerably whether or not the polar cap could be solid carbon dioxide. The observed partial pressure of CO_2 on Mars is roughly that expected to be in equilibrium with solid CO_2 in the polar caps. The most diagnostic is the infrared spectrometer–actually absorption bands for the CO_2 around 3 microns are only visible in a very dry experiment because they are normally shielded by very small amounts of water. Here is *prima facie* evidence for the existence of solid CO_2. The infrared radiometer and the spectrometer experiments both gave temperature information which has finally been brought more or less into agreement; both–near the polar area–reached temperatures of the order of 150°K. For several reasons that number is not a tenth of a percent number in energy but it is reasonably good, so basically, the CO_2 model was confirmed.

The infrared spectrometer people have attempted to infer the thickness of the cap, based on looking at essentially glare ice, CO_2, laboratory specimens. They note the difference in absorptivity between two different wave length bands and they try to use it as a measure of the thickness. Unfortunately this gives no information about the composition of the substrate. They did see some of this effect in their spectra and interpreted the cap as being a millimeter or a millimeter-and-a-half of frozen CO_2 on the surface.

They are seeing 1 or 1.5 mm of glare CO_2 rather clearly sitting on something else, presumably corn snow CO_2 with many reflection surfaces. The only discrepancy in the interpretation is that work in the laboratory has not simulated what would be expected from thick layers of CO_2 snow.

YOUNG: Can estimates be made on the basis of the retreat time on the polar cap?

MURRAY: Yes, the model that was published in 1966 gave its thickness as a meter or several meters, matched the recession rate, and the existence of a permanent north polar cap. This is explained. The new data are entirely consistent with frozen CO_2, the only discrepancy is the interpretation that the thickness observed is only a millimeter. Essentially that is the optical depth that is seen. This information doesn't help determine the total thickness. You can't expect to see through meters; only a little thin layer on the top will be transparent. Below is more granular material optically thick, and not seen.

LEOVY: The 1966 model of Leighton and Murray (Leighton and Murray 1966) has to be interpreted as an upper limit for the plausible annually deposited thickness. Several things could make the layer thinner than that model. I can conceive of nothing that would make it thicker.

MURRAY: Locally, with cratering, it could be thicker. This is speculative. Snow could be blown in, and if crater slopes are below that of the average for that latitude, permanent ice sheets may possibly be built up inside the crater. Some features that are seen in the polar cap, these enhanced central peaks [pointing], might indeed be that. Mariner '71 can tell us this. If little white patches inside craters are seen they are probably permanent ice sheets.

YOUNG: Water ice?

MURRAY: No, solid CO_2 ice.

YOUNG: Why couldn't water ice accumulate in protected places?

MURRAY: Let's review the water argument. You might also ask why water ice is not at the polar cap. The 1966 work argued that the composition is frozen CO_2 based on correspondence with pressure, retreat rate, and so forth. It clearly is not water.

The reason that the polar caps could not be water was known long before Mariner '67. Atmospheric water could not be transported. We know from observations that 15 to 30 micron columns of material go from a polar cap at one end of the planet to the other in about 90 days. Assuming a horrendous wind system, complete mixing and deposition of water at the other pole, I calculate that that is needed to get a millimeter of water laid down–or some very small amount to give a white appearance. Assuming 500 kilometer-hour winds for 90 days in the same direction, complete mixing of the column every 24 hours gives you less than a millimeter of water laid down. This is a small amount. The only way water can exist at the caps is if the water is coming in out of the surface at that point.

YOUNG: There must be some water in the cap–whatever water there is in the atmosphere will be frozen out.

MURRAY: Yes, but that is a tiny amount.

YOUNG: Couldn't that accumulate in the bottom of a crater where perhaps it might never melt?

MURRAY: It could accumulate up to a limit, which might be said for the north polar cap, too. As Leovy says, there can't have been much. Yet over 1 or 100 million years, there surely must have been a small amount, enough water to have produced a huge glacier at the north pole, which is the colder of the two poles. This also may have occurred in permanently shaded or nearly permanently shaded areas, too. This breaks down, however, because the cold pole of Mars reverses every 50,000 years since the orbit is eccentric. The cold pole of Mars is a product of the present orientation of the perihelian of Mars. Both are recessional with regard to the heat conditions every 50,000 years. Whatever net accumulation has taken place over 50,000 years at one pole is forced to migrate once every fourth time. Since every 50,000 years it reverses, systematic accumulations are only for that period of time, still for water, of the order of meters at most. There may well be meters of water in isolated areas. My guess is that everywhere it goes below the surface, water will be found.

The observations for Mariner 6 and 7 have consistently agreed that this cap is CO_2, at least millimeters thick, and likely to be centimeters to a fraction of a meter thick. That is no longer speculation but observation, meaning solid vapor equilibria have to obey those constraints. The behavior of water or organic materials must be thought of as in a U-tube, if you will, with a CO_2 cold trap in it as a vacuum system. Mars has a simpler atmosphere than was originally supposed—something like the moon mounted with a huge fire hydrant letting CO_2 shoot out for a while until it solidified and eventually formed a vapor. This model explains Mars well.

What I was hoping we would discuss was the carbon suboxide proposal from here on.

KAPLAN: In 1966 or so I asked you about evidence for sulfur dioxide or hydrogen sulfide in the cap regions. If substantial amounts of sulfur dioxide were in the ice caps would it be resolved spectroscopically?

MURRAY: According to the slide we saw before, the upper limit for sulfur dioxide is 2 parts per million.

OWEN: That is gaseous, mixed with the atmosphere, of course.

MURRAY: Is that 2 parts per million absolute or 2 parts per million of CO_2?

OWEN: Two parts per million in the atmosphere.

MURRAY: So, it is much lower than CO_2.

KAPLAN: Two parts per million of CO_2?

MURRAY: Yes, and a test that can be applied to any volatile is its vapor pressure at about 150°K. If in equilibrium with that it can be seen, it is an observable.

KAPLAN: So, atmospheric SO_2 is probably insignificant.

OWEN: It is not quite that simple. Applying this test to water, atmospheric water wouldn't be seen.

MURRAY: That's right.

OWEN: And we do see it.

MURRAY: The explanation is that they are out of equilibrium.

OWEN: Conditions are changing. It isn't just finding the vapor pressure—

MURRAY: But CO_2 is far more volatile than water. Solid water has an extremely low vapor pressure. It is a game of the vapor pressure in a temperature range from 150 to 200°K. CO_2 has a very high volatility, so for it there is essentially almost instantaneous equilibrium. Water is much less volatile so since it can be out of balance for about 10^4 years, there is no problem. How far out of balance can this be? Nothing, not even water, can be out of balance a billion years. If it weren't for the 50,000-year reversal phenomenon, there would be one huge glacier at the north pole of Mars. I think the problem is bracketed and we must live within the brackets to understand what the surface chemistry equilibrium is.

LEOVY: There may well be a serious problem with this equilibrium with water. This requires a highly inefficient mechanism for cold trapping it out of the polar cap every year.

OWEN: Right.

MURRAY: We have a very passive audience. After having given a rather extreme view of the planet everyone seems so filled up with lunch that its acceptance goes without dispute.

HOROWITZ: When will you discuss solid surfaces?

RICH: Let me raise something very unorthodox and a bit fantastic that Leslie and I have been whispering about: The prospects of life without nitrogen, or life in which nitrogen is like a trace element, there—but in small amounts. This living system would not build up regular polymers containing nitrogen because it is in short supply, but esters and polyesters chemically somewhat similar to amides could be used instead. This would be possible only in a somewhat acidic environment because esters and polyesters break down rapidly in alkali.

So, a point at issue is what the pH value of the Martian surface is. Does any evidence suggest it might be slightly acidic? This brings in the carbon suboxide questions: It might be, so to speak, an *amino acid,* a building block of Martian polymers.

Maybe now you will tell us about carbon suboxide.

HOROWITZ: I will tell you what I know, which isn't very much. I didn't really come prepared to talk about this, so if I make a mistake Toby or Leslie, please correct me. Carbon suboxide can be generated from carbon dioxide by ionizing radiation or short UV, and produced in the laboratory. It is a double anhydride of malonic acid. Dehydration of malonic acid gives carbon suboxide. It is a gas which polymerizes readily into a red polymer, the structure of which is as follows (see Smith et al. 1963).

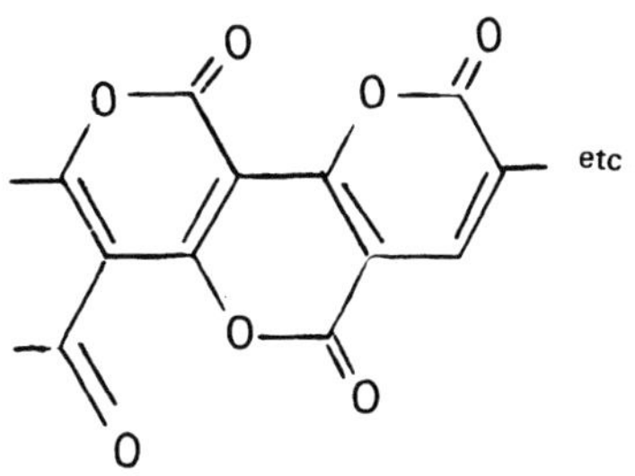

ORGEL: Is your statement that it can be made by discharges an hypothesis or from experimental results?

HOROWITZ: Experimental.

OWEN: Aren't several different polymers produced under different temperatures?

HOROWITZ: Yes, red and yellow polymers. I don't think the structural differences between the different polymers are known.

Carbon suboxide formation in the Martian atmosphere has often been suggested. Recently a paper in *Science* by some people from the University of Massachusetts proposed that carbon suboxide might cause the red coloraction of Mars. They did some spectral studies on the carbon suboxide polymer, and they found that the reflection spectra from one of the various polymers fits rather well the known reflections from Mars. They proposed that the polymers account for the red color and even suggested the production and degradation of the red polymer might account for the seasonal changes.

I compared the spectra that Plummer and Carson (1969) published for C_3O_2 with the reflections John Adams published in 1968. Adams compared the reflection spectra of Mars with reflections from basalt and the fit is much better than the C_3O_2 polymer. Adams found that if basalt is oxidized slightly by dipping it in 10 percent nitric acid so that it is stained with ferric oxide, it gives off reflections that fit the Martian spectrum nearly perfectly.

ORGEL: Didn't Dolphus claim that?

OWEN: That was limonite.

MURRAY: There never was really any evidence. From what is known about Mars, if limonite were really on the surface, this would be the most astonishing fact imaginable. It implies bog iron ore, a very wet iron.

ORGEL: Five years ago everyone was very serious about the limonite suggestion.

OWEN: Some people are still very serious about it.

MURRAY: That is part of the cultural lag.

KAPLAN: How about a dry bog, Bruce?

MURRAY: The point of interest that you are not being fair about is that there is no way to measure basalt's annual change of its spectral reflectivity. This is why this hypothesis deserves careful study.

ORGEL: Could it not be undergoing an oxidation-reduction equilibrium that is dependent on the availability of water?

MURRAY: At the temperatures and partial pressures of water we observe, I think it would be extremely unlikely.

ORGEL: But photochemical changes are well known—if you take ferrous iron and water in almost any form, you get ferric iron and hydrogen.

MURRAY: But the surfaces exposed to light on Mars cannot possibly have liquid water on them.

ORGEL: They can have tightly held, absorbed water.

MURRAY: Yes, but this is a one-way cycle. If the water is depleted there is no mechanism to replenish it.

ORGEL: When there is no water, hydrogen and ferric iron go back to water and ferrous iron.

MURRAY: Granted this happened once, how does it go again?

ORGEL: Add more water.

MURRAY: But the water is frozen.

MILLER: But there are 15 or so microns of H_2O in the atmosphere.

MURRAY: Are you suggesting this reaction would go with a frozen substance, at 200° Kelvin?

ORGEL: Photochemical reactions are almost, but not quite, independent of temperature.

SHOEMAKER: The problem is to keep alternating this environment from oxidizing to reducing.

ORGEL: Water-vapor pressure. If water is destroyed it must be remade from hydrogen.

MURRAY: All the water vapor—the amount needed—can come from some other place on the planet.

ORGEL: As long as the amount of water can be changed, I think it can be done.

MURRAY: How will this very small abundance of atmospheric water interact with the surface materials?

RICH: It only needs tight binding sites.

ORGEL: The intensity absorption is very high with these materials, so only a very narrow layer must be modified. If there were a thin layer on the surface with tight binding sites for water in the summer when water is available, it sticks on, water is decomposed to hydrogen, and the surface is oxidized. Subsequently, when no water is around, the process is reversed.

MURRAY: Why?

ORGEL: Because of the chemistry.

MURRAY: Does the ionization reverse it?

ORGEL: Ferrous iron in the presence of water is oxidized to ferric iron and ferric iron in the presence of hydrogen goes back to ferrous iron and water. Although I don't know if it would work this way on the Martian surface, these are well-known reactions.

SHOEMAKER: Ferrous goes to ferric very commonly on the earth.

ORGEL: In biology it goes both ways, ferrous iron goes to ferric and to ferrous rather easily in aqueous solution.

MILLER: Yes, that is correct.

MURRAY: I believe it. It is a well-known fact that things get rusty.

ORGEL: No, that is not the right analogy. We are talking about getting rusty in the sun.

MURRAY: It is the reversal I am interested in.

ORGEL: Ferric iron in water, in the absence of all the O_2 and put under an ultraviolet lamp, will slowly evolve oxygen and ferrous iron will be left. With ferrous iron in water, hydrogen will be evolved and ferric iron will result. A certain equilibrium will be reached depending on the amounts of water and light.

HOROWITZ: This is another point against carbon suboxide.

MILLER: Norm, how is the carbon suboxide polymer supposed to change color with seasons.

HOROWITZ: They didn't detail this in the paper.

MILLER: It would imply a breakdown of the polymers.

OWEN: Probably we shouldn't place too many hopes on that.

MURRAY: Beyond hope, I don't understand any plausible mechanism for what we observe, including this last one. The C_2O_3 must be considered plausible until proven otherwise; we must be careful because we are in desperate need of a mechanism to explain the observations.

MILLER: Are you going to discuss the decarboxylation?

HOROWITZ: No.

The first step of this carbon suboxide is the reaction to form it in the atmosphere before its polymerization. I wonder how stable C_2O_3 itself would be under the Martian UV fluxes. Would this remain as carbon suboxide in the solar UV?

MILLER: I don't know offhand. I assume it absorbs around 2,800 Å.

HOROWITZ: It has an absorption peak at 2,600.

MILLER: Just like DNA.

OWEN: Another problem is its formation in the atmosphere in the presence of water vapor. Is malonic acid produced immediately before the CO_2 is polymerized? This depends on concentration, once the malonic acid is formed it will be pretty difficult to get back to the CO_2.

HOROWITZ: That would decompose.

MILLER: The malonic acid, at least at normal temperatures, decarboxylates the CO_2 in acetic acid relatively rapidly.

HOROWITZ: Mars is dry enough to do the dehydration.

MILLER: No, malonic acid won't dehydrate under Martian conditions. You need P_2O_5, if that even works.

OWEN: At high temperatures.

MILLER: The C_3O_2 binds water very tightly and will never dehydrate by an equilibrium process.

MURRAY: What happens if it is exposed and zapped by the ultra-violet flux?

MILLER: In the dark eventually it will decarboxylate to acetic acid—which will also probably decarboxylate—and CO_2. I would be very surprised if it dehydrates. I have never seen an analogous photochemical reaction take off two waters at once and make C_3O_2. It will be made from CO_2.

HOROWITZ: There is probably some carbon suboxide on Mars.

RICH: What is the evidence for carbon suboxide on Mars?

HOROWITZ: None except that there is a lot of CO_2 on Mars and under Martian conditions carbon suboxide might be expected to be formed from CO_2.

MILLER: Have infrared lines been looked for?

OWEN: Yes, and they have not been found. This is in gas phase and says nothing about the polymer. I don't know if anybody has looked into the reflection spectra.

MURRAY: I think Pimentel was going to look with the mass spectrum.

OWEN: I am sure he will.

MURRAY: If it is plentiful enough to affect the color to the extent that Mar's seasonal variations do, it should be plentiful enough to detect by infrared-reflection spectra.

Going back to the oxidation reaction mentioned, what are some plausible mechanisms by which there could just be redox reactions occurring?

ORGEL: Let me put it on the board.

$$Fe^{3+} + OH^- \rightarrow Fe^{2+} + OH$$
$$2OH \rightarrow H_2O + \tfrac{1}{2} O_2$$
$$Fe^{2+} + H_2O \rightarrow Fe^{3+} + OH^- + H$$

KAPLAN: Leslie, isn't this a two-step process, which would be easier to envisage, if written as such? Isn't this a photochemical process reacting with water, and then the oxygen released reacts with the iron, and hydrogen escapes?

ORGEL: No. I don't want to go into the detailed chemistry, which involves electron transfer reactions. Ferrous iron where irradiated in solution gives ferric iron plus a hydrated electron. In time the hydrated electron breaks down.

KAPLAN: In the presence of light.

ORGEL: Of course. This is an exactly opposite reaction. An electron beam is taken off the water molecule; both reactions proceed only in the presence of light. The equilibrium between ferrous and ferric—at a given light intensity—depends upon the amount of water.

LEOVY: The so-called darkening wave presumably occurs when water is liberated—assuming that water is trapped in the polar cap and then moves from the cap toward the other pole. In the ground-based spectra is there evidence that this is what happens as far as the seasonal variations of water?

OWEN: No, there is no clear evidence yet as to which way the water is moving.

HOROWITZ: Doesn't Schorn's evidence show that there is no water in midwinter in the Martian atmosphere but as spring comes around it is seen over the polar cap?

OWEN: The difficulty is that his observations 2 years ago (Schorn et al. 1967) are in direct contradiction to those he made this year (Schorn et al. 1969). They show the water in the opposite hemisphere, and he is not sure which is right.

HOROWITZ: That's pretty bad.

OWEN: I think this year's observations are in the right direction; they follow what you said and are probably more reliable.

YOUNG: Saved again.

ORGEL: If anyone is interested this should be fairly easy to test. A mineral of the type claimed to match the Martian spectrum could be irradiated in the presence of more or less water.

HOROWITZ: Leslie, I was supposed to send you some basalt or something about 2 years ago—wasn't I?—and you were going to do this.

ORGEL: I don't know. Anyhow, I didn't do it.

MURRAY: I have the personal feeling that if we had had this discussion in 1964, people would have jumped up and down and would have been full of ideas about life on Mars and its significance. Today somehow we have orders of magnitude more data and a lot better perspective, yet it doesn't seem to produce even a ripple. I am curious about this.

RICH: What do you mean by a ripple?

MURRAY: People don't seem involved, it doesn't seem to mean much.

ORGEL: We are waiting for tonight's discussion to know if ideas of subsurface water are likely or acceptable.

HOROWITZ: Let's get to the history of water now.

MURRAY: Should we go on to the conclusions and speculations concerning surface morphology?

ORÓ: My personal feeling is that you are interested in a "carbon suboxide type of life" which doesn't interest biologists here and, on the other hand, you don't seem as interested in our kind of life.

MURRAY: O.K., that is fair. I am an observationalist. The variations in light and dark areas is seasonal or otherwise is a fact. We still have no good explanations, so when someone tosses one out I think carefully about it. Conversely, when someone tosses out a suggestion for which there are no relevant facts, I don't waste my time. This C_2O_3 stuff is relevant only because it purports to try—or its critics claim, maybe—to explain the known property of the planet.

ORÓ: The real reason I have not reacted is because I am having my siesta, but let me say that I fully agree we need work done on carbon suboxide. Kuiper (1970) proposed this many years ago for Venus and it was criticized; now it is brought up again for Mars. It is time we did something about it. Another area in which work is needed is in the area of nitrogen.

We argue about the low nitrogen content or its absence but practically no experimental work has been done in this respect.

RICH: Reflecting on the lack of nitrogen, I have done some work that I will tell you about. Are terrestrial ribosomes–which are busily manufacturing proteins, which are polyamides–capable of manufacturing polymers without nitrogen? They are. If an alpha hydroxy acid, suitably activated on a transfer RNA, is fed to the ribosome with an appropriate messenger, polyesters will be made. For example, with a polyuridylic acid messenger RNA, I have made polyphenyl lactic acid esters.

The ribosome clearly has the ability to make esters as well as amides. The reverse reaction, the hydrolysis of peptide bonds, is characterized by the fact that all proteases are also esterases. Chemically the system can function with either ester or amide linkages. Whether this provides enough information to fabricate not nitrogen-free life but life in which nitrogen is a rare element, it is hard to say.

MARGULIS: We certainly can't leave these astronomers with the impression that your transfer RNA and ribosomes aren't full of nitrogen.

RICH: They are sophisticated enough to know that. Organic chemists know that amides and esters are similar–now I think the biochemist also has a basis for this similarity. Whether that small fact is enough to make a big theory is hard to say.

MURRAY: Could nitrogen be in the role that phosphorus is in, in terrestrial life?

ORGEL: It is more like selenium.

ORÓ: I think this is a very good suggestion, to be given serious consideration. Even though Norm and I feel nitrogen is important, perhaps essentially everything can be done with carbon and oxygen. Life essentially depends on delocalization of electrons, and if this can be done with carbon and oxygen then nitrogen may not be needed. It will not be as sophisticated a living system as one that has more than one heteroatom, but he might be entirely right.

YOUNG: Since we don't know that there is no nitrogen, we don't even know if we have a nitrogen problem.

ORÓ: He is acting like you–just raising problems.

MILLER: He is showing that if there is a nitrogen problem, there may be another way around it. I think this is directed toward Bruce's pessimism about life on Mars.

MURRAY: I have great confidence that imaginative scientists will find a way around the problem.

KAPLAN: Was it decided this morning that if nitrogen is present it should be as nitrates in the soil? Was this the consensus?

MURRAY: There was no consensus.

KAPLAN: May we make an analogy with the lunar samples. There is something rather peculiar in the results of Carleton Moore in Tempe, Arizona. I think only he carried out an extensive nitrogen analysis and he found that the carbon/nitrogen ratio was roughly 1:1. Nitrogen was not there as a simple gas. He believes it is not a nitrogen gas because it is not volatilized off when heated to 200° or 300°.

Based on the reducing properties of the lunar material, he interpreted his data to signify that the nitrogen was in the form of ammonia: nitrogen-hydrogen. If the nitrogen content of the lunar rocks was high and if the ratio to carbon is roughly 1:1, and nitrogen is bound, could it not be bound as ammonia on Mars also? Would ammonia be unstable under Martian conditions?

OWEN: Yes, if exposed to CO_2.

MURRAY: If exposed to the atmosphere; it is dead so far as ionization is concerned.

KAPLAN: Why?

HOROWITZ: Ammonia begins to absorb UV at about 2,300Å.

MURRAY: It is pretty fragile.

KAPLAN: Would it be converted to nitrogen?

MURRAY: And volatilized very quickly.

HOROWITZ: Ian, the consensus this morning was that there would be as much as 5 percent nitrogen undetected in the Martian atmosphere. No nitrogen in the soil needs to be proposed. It still could all be in the atmosphere.

KAPLAN: Five percent of the total atmospheric pressure is still low by analogy.

HOROWITZ: No, it is exactly right in proportion to CO_2.

MURRAY: We better pick that up at cocktail time. We have walked that path without gaining on it three times.

SHOEMAKER: Something bothers me. Nitrogen need not be abundant in any sense; it could be present in very small quantities. Marine organisms can concentrate and precipitate out of sea water the elements they need by many orders of magnitude. Calcium fluoride and vanadium are good examples. Trace amounts of nitrogen in the Martian atmosphere may be plenty to sustain life.

MARGULIS: Organisms can go all the way through their life cycle with no N_2 in the gas phase.

HOROWITZ: This is academic. There is no way to get a planet with abundant CO_2 and no nitrogen. How can an atmosphere be produced with all that carbon and oxygen, without any nitrogen? Nitrogen is just as abundant in the sun and cosmically as oxygen and carbon. I don't believe there is any possibility that the planet Mars is devoid of nitrogen.

MURRAY: Unless formed by comets. Toby had a fractionation by formation of solids, solid ice, maybe some solid CO_2.

OWEN: Comets have nitrogen; they will bring in ammonia.

HOROWITZ: Cyanogen.

ORGEL: It won't make ice.

MURRAY: Let's discuss the Martian surface and, if possible, identify the most likely processes that may have formed the surface. This is like trying to infer the appearance of the beast from the coprolite. In the cratered terrains which presumably have been formed by impact processes, we can try to reconstruct the history of removal processes, too.*

First I will show some slides of the uncratered terrains and describe their implications, which are rather short-term. We think they are remains of processes that have occurred fairly recently in the history of Mars. I'll spend the rest of the time talking about the cratered terrains (Leighton and Murray 1971).

OWEN: For purposes of distinguishing cratered and uncratered terrain, can you tell us what fraction of the total surface area of the planet has now been photographed?

MURRAY: At the familiar kind of resolution where craters are seen clearly, about 10 to 20 percent, over quite a range of latitude.

The surface is entirely covered with craters except in two general areas. One is Hellas, which has nothing–I will show a mosaic photo and discuss it separately. Another area, at the same latitude as Hellas, does have craters so it is not a latitude effect. Hellas itself very significantly is a regionally distinct area.

In the second area which doesn't show well here we have a chaotic terrain, the other kind of noncratered terrain. We presume a high impacting rate and that craters at least a kilometer in size are formed frequently on Mars as the result of the bombardment of cometaries. The absence of craters therefore must be equated with recent removal processes. Whatever else has been going on, craters must have formed. There is no reason why craters would not form at some place. Therefore, the absence of craters means we are looking at something recent.

---

*For the illustrations referred to by Dr. Murray, see Volume 76, number 2, Special Papers on Mariners 6 and 7 in the *Journal of Geophysical Research,* 1971.

Other areas are heavily cratered, the dark areas are heavily cratered (Sharp et al. 1971, p. 331). It is extraordinary that the largest heavily cratered surfaces have topography 3 billion years or very old by anybody's definition. This implies that whatever happened in Hellas never happened in the heavily cratered zones because that terrain would have been removed. Not only are there recent processes that are operative on Mars but they are geographically confined and always have been so. Extraordinary.

MARGULIS: By implication, there are craters in the polar zone?

MURRAY: Yes. They can be seen here [pointing] (Sharp et al. p. 357).

OWEN: Would you estimate the percentage of uncratered terrain?

MURRAY: Ninety percent of Mars is cratered.

OWEN: So about 10 percent is uncratered?

MURRAY: Yes, about. We suspect Elysium, which is flat on the radar, might be this way. Since we ascribe a mysterious origin to Hellas and Elysium is similar, we ascribe it to Elysium, too.

SHOEMAKER: Being flat to radar isn't necessarily the same as smooth here [pointing]. This is a different scale of flatness.

MURRAY: Let's go on. This shows the so-called chaotic terrain (Cutts et al. 1971). The sun lighting is very high, about 45 or 50°. There are no shadows in any of these pictures; that is one of the reasons the relief looks so flat. The only light and dark variations associated with topographic forms are just due to slope variation.

Along this boundary is a jumbled-up area. That is the only way to describe it. It is up and down in some arbitrary rough form and few craters, if any, are visible.

Here is another sharp boundary and inside this jumbled-up terrain again. These appear to be incipient fractures of an erosional process, whatever it is, cutting out in front.

SHOEMAKER: What is the scale?

MURRAY: About 50 kilometers across. We can't measure the relief accurately but I estimate this floor must be several hundred meters from these edges. It is clearly younger than the surrounding terrain because it is cutting into it.

OWEN: Do the light to dark boundaries correspond to any classical boundaries?

MURRAY: Yes and no. In a general way we find correspondence, but in a detailed way we do not.

This chaotic terrain indicates on Mars the fingerprint, footprint, tailprint of some recent surface process that was able to erode the previous existing topography on a dramatic scale, like the Bad Lands of South Dakota. This process must have been in operation recently because there are no craters in these areas. Furthermore it is geologically confined. I have some speculations about this extraordinary unknown process that I'll talk about later.

This shows a mosaic of Hellas area showing the other kind of uncratered terrain.

KAPLAN: What is the relief?

MURRAY: We don't know exactly; my estimate is several hundred meters, at most a kilometer.

SHOEMAKER: It looks mottled there [pointing].

MURRAY: That is not real. The argument can be made that ground fog or haze obscures the craters–preventing us from seeing the surface. But the craters begin to diminish, where we are sure we are still seeing the surface. There are other indirect arguments but this is the best direct argument that we are looking at a barren surface.

KAPLAN: What is the size of those ridges?

MURRAY: About 40 or 50 kilometers in this dimension.

STROMINGER: Is that feature standing up or going down?

MURRAY: Standing up. That dark lineation forms a structural boundary here. The explanation of Hellas must account for a very ancient heavily-cratered surface. A debris layer that might have resulted from a recent impact to form Hellas is not seen. Whatever formed the Hellas basin must have happened a long time ago, probably when the crust itself formed. If it formed later we should see remnants of the secondary ejecta, outside here [pointing].

Comparable to the origin of the lunar maria the origin of the basin itself must be back very early in history. However, in this barren region some process either scrapes out craters at a high rate or prevents them from being formed.

SHOEMAKER: Or fills them.

MURRAY: Yes, that is what I mean, by *scraping out:* Their topographic features are removed by erosion or sedimentation. This probably is a high area. By the appearance of these pictures it is implied but not proven that it drops off sharply here. This portion of Hellas is lower than the surrounding area, there is a physical scarp across there. This edge of Hellas is lower.

RICH: Do you base this conclusion on the shadows?

MURRAY: No, on the inverse spectrometer looking at the CO_2 width absorption line. Since the line is partially saturated, the height of the surface is mapped as an increase or decrease in equivalent widths on the lines.

RICH: How deep is the depression?

MURRAY: Believing the numbers, about 3 kilometers across there. There is a tough problem which we haven't entirely solved yet, but qualitatively it is a significant height difference, at least a kilometer and probably more. Estimating indirectly the amount of slope to get this brightness change indicates pretty good differences in height.

OWEN: Will you review briefly the ground-based information about Hellas?

MURRAY: Yes, it is worthwhile discussing Hellas and chaotic terrain, but rather than just looking at pictures let's try to see what it means.

The cratered terrain represents a much more ancient phenomenon so let us separate this and talk more about Hellas. If we accept the observation that 90 to 99 percent of the surface is without craters, we are forced to conclude that either current surface processes remove craters by burial, erosion, or some combination–or craters do not form there. On impact, the material is different and does not support craters. This seems wild, but if there were great thicknesses of ice on Mars conceivably the pseudoplastic kind of situation could be there–a big hole blown in it just gradually fills in and pretty soon is not seen anymore.

SODERBLOM: Or substantial numbers of crater-forming materials don't penetrate the atmosphere.

MURRAY: Right, like the Tunguska meteorite episode on the earth. Presumably a comet hit and put about 10 megatons of energy into the atmosphere but none onto the surface. A blast wave was created but it did not make a crater. This event, involving a large Siberian meteorite in 1908, is reasonably well known and one wonders at what altitude it happened. It led us to wonder about the role of comets hitting the earth. In the case of the moon–with no atmosphere–the comet hits the moon surface, and the impact phenomenon is similar to an asteroid. There is enough atmosphere on Mars to make a significant difference. Presumably the amount of fractionation energy into a blast wave or a surface-excavation phenomenon is related to the height in the atmosphere.

We hypothesized that Hellas was just the lowest place in the whole planet and the great majority of the impact flux was comets. But this isn't the answer; it is worrisome that we don't really know the fraction of asteroidal versus cometary impact even on the moon. It could vary over quite a large range, and the role of cometary impact on Mars could be very different from the asteroidal impact.

SHOEMAKER: One way of eliminating this hypothesis is to determine the altitude at which cometary or fragile objects break up in the earth's atmosphere. They broke up at about 5 kilometers. Such objects clearly have penetrated to Hellas unless Hellas is covered with an atmosphere comparable to that of 5 kilometers above sea level on the earth. Almost any object that is capable of making a crater about 2 kilometers in diameter on the earth's surface will get through. The issue is the dynamic pressure which will obviously shear and break up.

SODERBLOM: Is it a compression by deceleration?

SHOEMAKER: Yes, essentially.

SODERBLOM: Does this mean a larger body stands less chance of getting through the atmosphere than a smaller body?

SHOEMAKER: No, because the stresses are spread out over a greater distance.

SODERBLOM: Any material at the bottom or front edge supports the deceleration of all material behind it. This is going to increase the quenching of the diameter of the object.

MURRAY: It is a heat-shield problem. It will only build low entry coefficient heat shields of a certain size.

ORÓ: In summary this means that on Mars there is not too much difference between meteoritic and cometary impact.

MURRAY: I am not sure I accept that conclusion.

ORÓ: Let's say 100 times less.

MURRAY: We don't know whether a comet is a single object or a collection of objects. On the moon there is no difference; the same impact is the same whether caused by a bunch of BB's or objects the size of this building. Only the total amount of mass and the velocity at which it comes in matter.

YOUNG: Is there any chance for a glancing blow on Mars? Could it skip through the atmosphere?

MURRAY: Only if it comes in at a low-level angle, but again the orbital conditions are different. The entry velocity may be different on Mars than on the moon.

Our conclusion so far is that the Hellas basin, that big structural feature, must have been formed early. We don't see how it could have formed last week and then gotten rid of the craters without leaving telltale signatures. We think the basin formed early, and that it is an ancient feature.

SHOEMAKER: Why couldn't the basin just have been formed by subsidence? Why does it have to be impact?

SODERBLOM: Because the craters on the edge of the basins are not faulted by the structures that run through them.

SHOEMAKER: But there are some beautiful layered structures.

SODERBLOM: Linear structures in the scarp zones are overlaid by craters that are not distorted. The craters have to fall after the scarps are formed.

SHOEMAKER: It was clear that the scarps are old from the picture?

SODERBLOM: It isn't easy to see.

MURRAY: The basin or the ground structural feature, which at least on one side is low compared to the surroundings, is very ancient but a current process prevents craters from remaining there. We don't know what the process is. My colleague at Cal Tech, Bob Sharp, plays around with what he calls a popcorn model, but all we can say is that something is going on. We talked about the cratered terrain but at some places some process of a different magnitude occurred at different times, and clearly there are geographically confined erosional processes of very large magnitude on Mars which are not understood.

SHOEMAKER: Why erosion?

ORGEL: Why can't sand be filling craters in?

MURRAY: Coming from where?

SHOEMAKER: Maybe lava burped from the inside of Mars.

MURRAY: That means it is burping right now.

We can suggest special materials at Hellas, unlike elsewhere, or special processes, or some combination of both, but any explanation is ad hoc and fraught with difficulties.

RICH: Could some giant ancient-impact crater fracture the mantle to give access to a slow lava leak?

MURRAY: This is one idea of the origin of the lunar maria.

That is a lot of lava right now. There are other difficult-to-understand observations, too; for example, this area tends to brighten and is apparently the brightest area on the planet at various times. Frequently it is mistaken for the south polar cap. Many things about these bright areas, which appear and disappear, we do not understand.

RICH: Do you simply get good reflection because of the smoothness?

OWEN: The reflection is not good all of the time.

MURRAY: No, it is not specular, not a phase-function effect. The albedo really goes up.

RICH: Every time it floods with lava?

MURRAY: Yes, every time it floods. Some things we have thought about would surprise you. But not lava. If it is to be made of something that will not show effects of impact it should be mostly ice, like a glacier with a thin layer of dust. It flows nicely.

LEOVY: You sound like a biologist.

MURRAY: Confronted with an inexplicable good observation I get serious, just as with the light and dark markings.

KAPLAN: By terrestrial analogy, this is an area of sedimentation, not erosion. Where does *erosion* come from?

MURRAY: Sedimentary transport must be occurring no matter where the material is coming from. Something removes that topography, either by scraping it away or burying it, or a combination of the two.

OWEN: On your model, what is your time scale estimate for this process?

MURRAY: My estimate is very rough. Perhaps at least one detectable crater should have formed there in the last 100,000 years.

OWEN: As recent as that?

MURRAY: Yes.

ORGEL: What is the obstacle to the simple-minded theory that some material fills the craters in quickly?

MURRAY: *One,* what material? *Two,* why only at Hellas and no place else? On the popcorn theory, we ad hoc and pretend there is such a material, but that is not an explanation. We don't know what the material is, why it should be there, or whether it is plausible.

KAPLAN: By earth analogy, that is not too illogical. If Hellas is a tectonically active depression basin, it might also have flows. Volcanic activity here is localized, not spread out equally throughout the entire surface of the earth.

MURRAY: May I postpone that? The cratered terrain indicates there has been no tectonism on Mars for about 4.5 million years. It therefore is hard to believe extensive activity on Hellas never even interrupted the rest of the surface.

KAPLAN: What was the elevation of these ridges? Were they about 40 kilometers long?

MURRAY: Elevations can't be measured directly in those pictures without better photometry than we presently have. Any individual ridge probably is not higher than a kilometer.

KAPLAN: If there are active sediment transport systems, it is possible giant sea dunes were formed there? Dunes on the earth may be many kilometers long.

LEOVY: Craters intersect these ridges. There are no craters in the center of Hellas and yet craters intersect the ridges. There is an interesting wind erosion possibility. If Hellas is several kilometers lower and therefore the mean surface pressure increases from 5 mb to 10 mb, the threshold wind that is needed to lift ordinary 100μ range dust particles is lowered from about 50 meters per second to about 25 meters per second. This decrease in threshold wind needed may make a big difference on the surface of Mars.

MURRAY: It might be a great trap.

The chaotic terrain is recent, by the same argument, and geographically confined, but an active erosion or modification process is involved. There does seem to be a very weak correlation with the chaotic terrains and the physically low areas, as determined by radar. Sharp has suggested that Mars has been heating up internally and only now are we beginning to get active tectonic processes or the geothermal gradient has moved up high enough to melt permafrost and cause collapsed structures. He is suggesting this very interesting possibility—and quite pertinent to the question of life on Mars—that we are beginning to see Mars becoming an active planet.

Although it sounds like special pleading, it is consistent with the observation of this phenomenon only in restricted areas. It is extremely hard to imagine surface-process mechanisms restricted to some small area over billions of years. So Mars may have been a cold, dead planet that now is warming up and something is beginning to happen.

SCHOPF: A cold, geologically-dead planet.

MURRAY: It has been a cold planet that has not differentiated.

SCHOPF: Do you think we should land in Hellas?

MURRAY: Fortunately, we don't have to decide that until we see some '71 pictures.

SCHOPF: What is the diameter of the Hellas basin?

MURRAY: About 1,500 kilometers.

SHOEMAKER: Why can't the *popcorn* be ice? What is the latitude?

MURRAY: It may be ice. We can talk about that.

MARTIAN CRATERS, WEATHERING, AND THE MARTIAN ATMOSPHERE

The cratered terrain question is conveniently broken into two categories: the implications of 1. the very large craters; and 2. of the smaller craters. The two are not in equilibrium with an impacting surface. If we were looking at an equilibrium surface, many more small craters should be seen—since the big craters make secondaries, small craters are formed. For each large impact there are many more small ones because of the greater number of smaller impacts. We are going to conclude that the present surface is not the result of a single event or single kinds of events in time, but that there were one or more episodes of history represented in it. The existence of these very large craters in a nearly saturated condition—that is, nearly overlapping on one another—suggests that topography is a remnant of the original planet formation episode, the primordial surface that is comparable to the lunar uplands. The smaller craters, their absence as well as their presence, are considered indicators of events that transpired after the original formation. I will try to develop these ideas.

If these can be demonstrated, some conclusions can be drawn about the ancient history of Mars just from the fact that the primordial surface has survived both internal and external modification processes. In the case of the small craters we conclude rather mysterious events must have occurred in Mars' history. The next slides show both large and small crater frequency.

[Pointing] This is a mosaic showing Diplionius Regio, the Bay of Sinus, and Meridiani Sinus. This light to dark boundary is the Diplionius Regio. The light and dark patterns–related to the way these particular pictures have been handled–are a bit misleading. These are resolution frames from the overlapping areas.

This shows a representative crater terrain surface, about 2,500 kilometers across. The high resolution is about 100 kilometers. The population of small craters is low compared to the population of large craters. It looks funny, strange, like two different surfaces. This was immediately apparent, receiving these pictures in real time. This low concentration of small craters is a principal factor about the planet. The large craters all appear to be flat-bottomed forms; they appear filled in. No secondaries, no rays are visible from these craters. Clearly the large craters at least have been superficially modified. The smaller craters are round and bowl-shaped, they resemble primary-impact craters on the lunar maria. Even this 50-kilometers-across crater is flat-bottomed. There is detail on this crater's side, but it clearly has been modified: Something has been filled in since it was formed.

Other features are visible. Many small ridges are similar to but not really the same as lunar ridges.

RICH: Very few central projections.

MURRAY: Yes. Looking at these figures in different views I will show a slight morphologic difference between the light and dark areas. In fact there is very little significant morphological difference between the light and dark areas. They are both cratered.

The craters here tend to be brightly rimmed which makes them more conspicuous at high lighting. That is true in all the dark areas. There is an albedo effect, presumably superficial and tied into the change in appearance with seasonal variation.

This shows the terminator. The light and dark boundary runs here [continuing his pointing], a pattern of cracks is quite visible. A pattern exists; there appear to be colliginal cracks. Remember that this is 90 kilometers across so they are big features–with a crack-like pattern which manifests itself right in this range in here. They seem to mimic a colliginal pattern in the crust material itself.

SHELESNYAK: Are they similar to those in Arctic terrain?

MURRAY: I doubt it. This is a very large-scale phenomenon. There really are cracks there. They also appear only on the dark side and not on the light side of this boundary. Whatever the explanation there does appear to be a morphologic difference between the light and dark areas, yet this is the only other difference we can see between the light and dark areas in the pictures.

SODERBLOM: The cracks can be seen from back here–I see about four.

MURRAY: The term *rill* has been used for the moon but since *rill* has a specific meaning I hesitate to use it.

One sees just more of this terrain. Apparently it is not really a saturated surface–everywhere you look you don't necessarily see a large crater bottom. The elevation angle of the sun is about 27 or 30° at the middle of this picture, and a picture of the lunar uplands taken at the same sun-to-earth angle looks quite flat: It has no shadow. We are going to compare the crater frequency in this picture with that in the lunar uplands.

This shows the crater abundance in the conventional way that crater statistics are plotted. This is an accumulated curve, based on a unit area of 10^6 kilometers. The statistics back up the pictures; there are far too few craters in this range at the smaller end.

I am discussing these observations quickly because of their relevance to the interpretations with which you may want to argue.*

First we have the large craters–20 to 200-300 km. Forgetting Mars, we can ask how old the upland surface of the moon is observed to be. If we knew the impact-flux rate and could count the craters, we could estimate a minimum age, assuming that the craters that were formed have been retained.

How long did it all take to happen? On the basis of the flux rate that is presently observed on the earth or the flux rate that is implied by the Apollo 11 and 12 samples – where we can calibrate because we know the age of the rock from which those craters were formed the uplands are older than the history of the solar system. Therefore, the flux rate must have been much higher in the distant past.

---

*For the observations mentioned see *J. Geophysical Research,* vol. 76, p. 293ff, 1971.

This should not come as any great shock. That is, we usually call this the planetesimal phase, presumably the final formation of the lunar surface itself. The Apollo ages work out very nicely in that regard.

To understand the surface history of the moon this heavily cratered surface must be associated with the planetesimal phase, the final accretion phase of the planet itself. In a general way we can understand what has happened since. Now we see a similarly heavily cratered surface and so we play the game for Mars. This was first done at Mariner 4. We assume that all the impacting objects on the lunar surface come from the asteroidal belt. If they are all asteroidal, we estimate—on the basis of Monte Carlo and other techniques—what the ratio of impacts would be on the moon when compared to Mars. We get a factor of 20-25 higher. Assuming all impacting objects were asteroidal, and none were comets, we find for every asteroid that impacted the moon in the last 3 or 4 billion years, there must have been 20 or 25 that impacted Mars.

If comets were involved, the ratio is less. The comets can be out of the plane ecliptic lowering the selection factor between Mars and the moon. It is hard to imagine why Mars was impacted at a higher rate than 20-25 above that of the moon. Using the old data from Mariner 4, the estimated age turns out to be of the order of 2.5 billion years. Since we have better pictures and we can count more craters, the crater-frequency distribution itself has gone up observationally. This raises the lower limit to about 5 billion years, assuming only asteroidal impacting objects are involved.

MILLER: Didn't Jim Arnold (Anders and Arnold 1965) calculate an age of 300 million years from the ratio of the number of expected impacts?

MURRAY: He used irrelevant information about the impact flux on the lunar surface. The Arnold calculation, taken with the Apollo 8 dates, may have been off by a factor of 2.5, but it was off an already extreme model. Only objects that would have hit the surface of the moon since the formation of Mars came from the asteroidal belt by that process.

If we accept Gene Shoemaker's—and other people's—idea that comets are the principal source of impacting, we are much closer to unity for the moon and Mars.

MILLER: Are you saying, given the numbers and scaling factors and so forth, that there is absolutely no way to make the craters on Mars more recent?

MURRAY: Yes. Arnold's—and others—original calculations took extreme cases, and even with the worst case 5 billion years is needed to form the surface. Our conclusion is that the surface is probably primordial.

By this kind of reasoning, the probability that the large crater surface is primordial has increased to about 90 percent, significantly higher than it used to be in the minds of man. Looking at the various sources of errors and how they could be underestimated, the most likely result still is that the large cratered Martian surface is primordial and synchronous with the lunar uplands. It is exactly the same kind of surface, and the only time there was such a high impacting rate was during primordial times. The Apollo 11 data yielded an important result because the absolute age of that surface was uncertain before those age dates came in. That whole realm of uncertainty has been removed and I don't think it is a negotiable number.

OWEN: Do you mean that if a substantial number of the old lunar craters were contributed to by objects in the earth-moon vicinity, and therefore they have nothing to do with the asteroids or Mars, that this tends to drive the age of the Martian terrain back?

SHOEMAKER: Let me try the argument again. Tranquillitatis is not a saturated surface. There is a saturated surface, the big craters in the Highlands, where the real cratering rate cannot be determined. Only certain limiting values can be calculated but Mare Tranquillitatis is still accumulating craters, but we didn't know its age. In fact I lost a bottle of champagne on a bet with Wasserburg about it. From the present rates at which objects are probably entering the earth's atmosphere, and the present spacing of Apollo group asteroids, I guess the surfaces at Mare Tranquillitatis to be as young as half a billion years, but it turns out to be almost as old as the moon. Its age is about 4 billion years, and it is hard to understand how the cratering rate could be less on Mars. It might be greater, but there are few reasons to expect it to be significantly less on Mars. Now the crater abundances on Mars with these larger craters exceed the numbers of craters on Mare Tranquillitatis.

MURRAY: By a very large amount.

SHOEMAKER: Yes. This leads to the conclusion of an age of at least Mare Tranquillitatis.

MURRAY: But if the extreme cases aren't used and we assume comets are half or more of the crater-forming objects, automatically ludicrous ages are calculated for the Martian surface. This reasoning leads us to high probabilities that this is a primordial surface. Only a extraordinarily favorable alignment of all the possible errors would exclude this but, of course, we can't ever be certain. There is a reasonable probability now between *likely* to *very, very likely*, depending on how you assess these possible hypothetical uncertainties, that the large cratered surface really is primordial.

MILLER: Accounting for all these factors, what is the probability that the surface is only 4 billion years old?

MURRAY: I think it is very low. The spectrum is going to range, the probability that it can be only 4 billion years old is not more than 10 or 1 percent–in that range–and the median value is certainly primordial.

The new data has led to much older estimates of the lunar surface. The flux rate is lower than previously supposed, therefore it is much more difficult to hypothesize something different at Mars than on the Moon. Whatever the probabilities were before, they are much greater now that these surfaces are primordial.

HOROWITZ: Has Arnold discussed this since the Mariner '69 data came in?

MURRAY: Not in print.

MILLER: Not to my knowledge.

ORGEL: Supposing the calculation was done naively without prejudices and a high figure came out, what would it be?

MURRAY: With the flux rates the same?

ORGEL: Whatever you think best.

MURRAY: Probably about 15 or 20 billion years.

SODERBLOM: The strong consideration is not the absolute flux rates or their levels, but the spectral distribution of material this size within those flux rates. The distribution—the slope of the curve, if you will—on the Mare has not changed in the last 4 billion years because the accumulations, although at different levels for different maria, have the same slope. The distribution that formed the uplands is different because there are many fewer small bodies compared to large bodies. That kind of surface directly correlates with Mars. The observed distribution of large craters must have been formed by the same kind of distribution where there was a small number of small bodies compared with large bodies.

If the two surfaces formed at different times, it would be odd that the lunar maria do not show the effects of this smaller distribution that was current on Mars 3 billion years ago.

MURRAY: This contradicts the previous model that said the asteroidal influx was for both. But he is saying the crater populations are different, forcing us to explain both the much higher factor for Mars and also why we see a different population than the moon saw in the same period of time.

RICH: Is there a correlation in terms of size?

MURRAY: Yes, size versus population.

SODERBLOM: The correlation suggests these had to be formed at the same time if the spectral distribution of masses is fairly even.

RICH: If a moderate weathering and covering is assumed on Mars the small craters would disappear first, so that the population would always be biased. Isn't this the explanation for different populations?

MURRAY: This consideration, to some extent, is built into the discussion of the large crater population. We are interested in the oldest regions—we haven't seen the entire Martian surface. If more cratered areas are found, the age will be driven further up. If those same areas photographed under better lighting show more craters, age goes up. Nothing makes it go down. The probability that at least some features of Mars are primordial is very great.

ORÓ: What happens to the age if you assume the impacts are all asteroidal or all cometary?

MURRAY: We have already assumed the all-asteroidal impact; with different kinds of asteroids, this gives the 5 billion age. If comets or any objects other than asteroids are involved, the age is automatically driven up. The comets will tend to be more similar on the moon and Mars, and bodies swept in late in the vicinity of the earth-moon system presumably would have been higher on the moon than Mars.

SHOEMAKER: Another point that may not have been clear in what Larry Soderblom was saying is that the observed distribution for the large craters, both on the moon and on Mars, is essentially a steady state distribution. Since the surface is saturated with large craters, we are not looking at a total record of cratering, but simply at a steady state population of craters. Therefore any ages calculated from large crater abundance are always minimum ages.

MURRAY: Let's move on. In terms of the biological implications, there is this high probability—my estimate is well above 90 percent—of primordial terrain on Mars, and the consequences need to be thought out. There are people like Carl Sagan who can constantly figure out some way to thread the needle to say there is only 70 percent probability, but it is not 10. This is not as uncertain as it used to be.

If we accept—in part or entirely—that this is primordial topography of these highly cratered terrains, then two geological statements may be made. I will state them and then give the reasons. The first statement is that no internal tectonism has occurred. No crustal movements that tend to destroy older features and make newer features, comparable to those on the earth, have occurred in this time period.

OWEN: At least not on a widespread basis, but doesn't that contradict Sharp?

MURRAY: No, he has only considered Hellas, not the primordial terrain. This is important—contradiction to older ideas of Mars being an earth-like planet. It may have been earth-like originally and later became the barren desert it is now. The assumption of some kind of internal differentiation and formation of an earth-like atmosphere is all dissipated. If we are observing primordial terrain, tectonism and differentiation never occurred.

SHOEMAKER: Or if it differentiated thoroughly, it must have been essentially at time zero.

MURRAY: Right. There has been no significant igneous history over huge areas of the planet, no island arc formation, none of the kinds of phenomena that have played an important role in the evolution of the earth.

Even if it is only 4.5 billion years of age, it is important to understand that it is a geologically dead planet.

MILLER: Geologically dead?

MURRAY: Right. Many people couple the situation needed for biology and geological phenomena like volcanoes and hot springs, and so forth. The Martian surface does not have current igneous activity or—

MILLER: Hot springs and volcanoes are not considered necessary for the origin and evolution of life in certain quarters.

MURRAY: True, but again this means the total spectrum of possibilities is getting fenced in a bit more.

OWEN: What do you mean by *geologically dead?* You have discussed now two different kinds of Martian terrain which have formed recently and unless the hand of God is involved, presumably some internal or surficial activity has occurred.

MURRAY: It is certainly true that the moon has an upland surface without plate tectonism and other kinds of history, and Mars has a considerable igneous history over the early billion-year portion. If Mars had a mare, which we haven't been able to find yet, a pristine surface may have persisted.

SHOEMAKER: Perhaps Hellas.

MURRAY: Not a mare. It has to be made of ice or something to be a mare.

SCHOPF: By analogous argument, would you say that the moon has been dead since the time of the Sea of Tranquility?

MURRAY: Yes. It is widely agreed that the moon has not undergone any planet-wide differentiation since then.

SCHOPF: If rocks dated 2.5 billion years were found from Apollo 12—

MURRAY: They were, but the chemical differentiation implied by the lunar rocks–as I understand it–assumed differentiation was an early event.

SHOEMAKER: We may be seeing the results of minor later activity.

MURRAY: Right.

SCHOPF: However, the moon is not really a dead geologic planet since crustal movement, differentiation, and apparently igneous activity of some type has occurred subsequent to its formation and at least as recently as half the age of the planet.

SHOEMAKER: Probably earlier. The correlation of crater densities on the various maria can predict that some happened half a billion years—

MURRAY: Whatever episode it took to outgas a lunar atmosphere happened very early in the moon's history.

MARGULIS: Geologically dead means—

MURRAY: The internal tectonism that would have accompanied such a dramatic event as differentiation and exhalation of an atmosphere must have happened before the uplands were formed or concurrently with them.

ORÓ: Can this be solved by saying *near* dead, like neofossil?

MURRAY: I am trying to get at the problem of the atmosphere with a two-pronged attack. The softer and weaker argument concerns absence of obvious tectonism; the other is that the existence of an earth-like atmosphere would have erased those features.

ORÓ: Is this valid if you are referring to the atmosphere? Could the moon have retained a significant part of the atmosphere?

OWEN: No.

MURRAY: No. According to current ideas of atmosphere formation the cataclysmic internal process would leave a surface record. Some kind of crustal structure that certainly disturbed the previously existing topography must have occurred, and different from the relatively local mare igneous activity.

SCHOPF: Do you mean that separation is required to get an atmosphere formed?

MURRAY: An atmosphere of the kind necessary for Mars would have had to have happened at least 4.5 billion years ago. If it is continual accretion, we couldn't have had it anyway. We know there hasn't been one since then. This is a soft argument that points in the same direction.

MILLER: Do you mean that in order to have an atmosphere—so far undefined—outgassing from the interior of the planet must have occurred, which implies finding some sort of tectonic activity?

MURRAY: Yes, the big bang or big burp model is essentially what I mean. The continuous creation or activity extended over 4 or 5 million years is already ruled out.

MILLER: Even though many people have accepted Rubey's (1955) argument that the atmosphere of the earth, including the oceans, came from outgoing of the interior, it is not obvious that it is correct. Those who know more about this than I do say many assumptions must be made.

MURRAY: The atmosphere and oceans either formed in a rather brief period of time or over a very extended period of time.

MILLER: I am saying it didn't have to outgas.

SHOEMAKER: Do you want a residual atmosphere?

MILLER: Yes.

SHOEMAKER: Rubey's argument concerns the low abundance of rare gases, and it is a very hard argument.

MILLER: Oh, yes. It is true the methane, ammonia, and dust cloud did not come in proportion to the silicate, but nitrogen and ammonia and organic carbon or methane can be retained in other ways. You can have a substantial residual atmosphere without the rare gases. The methane or carbon can be held in organic compounds, and in the case of the earth much of it probably was held that way because the ratio of the masses of water versus xenon doesn't explain the abundance of carbon on earth.

I have always thought the earth's primitive atmosphere was residual from the dust cloud and only in part that it came from degassing the interior. I don't know if that is even demanded.

SHOEMAKER: I don't understand why the ratio of heavy noble gases to nitrogen is so low. That is the key argument.

MILLER: Let's go to the figures. There is 1 part of 10^{11} of the neon; there is 1 part of 10^{6} of xenon; there is 1 part of 60,000 of the water in the primitive dust cloud and, I guess, 1 part in 10^{5} or whatever of carbon. This doesn't go as the molecular weight. Most of the nitrogen, water, and carbon was lost from the primitive dust cloud. Not all that amount was lost, some was held in some way.

HOROWITZ: As solid material.

MILLER: As solid material. Ammonium salts, hydrated minerals, and organic material. If there is a process for getting this on the planet–the collisions and so forth–without burying it completely, a substantial atmosphere can remain.

OWEN: But how do you selectively get rid of the noble gases?

MILLER: By escape from small particles. The rare gases escape from objects of low gravitational field, the standard explanation.

HOROWITZ: Stanley, you are talking about the theory everyone has accepted.

ORGEL: I think Stanley is making a slight modification. The standard explanation is that the whole earth is formed in a great blob and heated up in the middle, and then outgassed.

SHOEMAKER: There are two alternates: you are degassing from planetesimal, the planetesimal hypothesis.

MILLER: Yes. It is reasonable. The interior does not have to be degassed a la Rubey (1955) and still there might have been a decent primitive atmosphere to make organic compounds.

SHOEMAKER: This depends on what you call a primitive atmosphere; you are liberating atmosphere as you accrete.

OWEN: How do you explain the accumulation of argon 40?

MILLER: In large quantity, relative to argon 36, it must be done by potassium decay.

OWEN: But then how does the argon get out?

MURRAY: How do you account for the present argon isotope ratios on the earth?

MILLER: Obviously it is degassed.

MURRAY: If the argon has been degassed, why haven't the rest of the constituents come from the inside?

MILLER: I didn't say that none was degassed. I think some of the nitrogen, carbon, water was degassed from the interior, but it is not clear that it all was.

SHOEMAKER: This hypothesis is straightforward: the atmosphere is liberated as the planetesimals accrete. As it heats up the atmosphere is made early, the argon doesn't come out yet because it isn't available; it has to be sweated off the top.

MILLER: Right. A substantial primitive atmosphere–the reduced gas to make organic compounds–may have been followed by degassing of a certain fraction from the interior of the earth. Not necessarily 95 percent was subsequently degassed. Half of the earth's atmosphere, including water may have been present initially and the remainder subsequently degassed. In the case of Mars it could have been a higher ratio of gas coming from the initial event.

MURRAY: I can't argue against that, but if an internal origin for the Mars atmosphere is required it was formed early. An early atmosphere from the interior is very close to a "big burp hypothesis"–a little burp, anyway. This has tectonic implications which are not apparent.

I realize this is a weak statement. But here is a stronger one to attack the question you raised. If the surface is primordial as we said, it has never been eroded enough to remove those features. In the aqueous atmosphere, with enough water vapor and total pressure to permit water in the liquid phase, water is the ubiquitous cause of erosion, making the atmosphere extremely effective.

Various estimates have been made of how long unrenewed topography will last on the surface of the earth. How long will an impact structure last in the event of erosion? An upper limit number in the literature is some 10^8 years. The real number for any type of topography is probably much less.

MILLER: I am not a geologist and I don't know much about the mechanisms of erosion but, if there is a small amount of water but no flowing streams or mud slides, would the evidence of impact last much longer?

MURRAY: Either there is liquid water or not—

SHOEMAKER: But does rain fall or not?

MURRAY: This question is whether or not enough moisture is present for chemical weathering to occur.

KAPLAN: Chemical weathering also demands removal of the products.

MURRAY: On Mars now there is enough atmosphere to blow stuff around, but our problem is understanding the present weathering.

MILLER: Is chemical weathering via water and dissolved carbonic acid really this limiting?

MURRAY: Without liquid and especially without the very important aid due to freeze-thaw action, the weathering is strictly mechanical. It is much slower. The weathering is not due to streams and rainfall so much as to the moisture present, stable, free liquid water.

SHOEMAKER: You need moisture.

The weathering depends on transport of water. There are very dry desert regions on the earth where erosion now is extremely slow. To weather, water must be transported.

MURRAY: If any large bodies of water exist, if atmospheric conditions are such that a water is stable and does not immediately evaporate in any one place on the planet, this must be true all over the planet. The relative humidity must be high enough at that pressure that water either doesn't evaporate or, if it evaporates, it rains or water flows at that spot. On geological or even biological time scales, any stationary amount of water goes somewhere. It is always evaporating. But the question is if it is coming back again, is there a water cycle? If it is not lost there must be a replenishment of liquid water, call it rain or whatever.

MILLER: Suppose the temperature conditions are such that water is always frozen or in the form of snows, except in the summer when it melts and evaporates. Is that rain, will that cause the erosion needed?

MURRAY: Do you mean the whole planet is frozen except at the tropical latitudes?

MILLER: When the water ice cap melts to form puddles, it evaporates over to the opposite pole.

MURRAY: This implies the triple point of water is maintained at that particular location. The entire atmosphere essentially becomes saturated.

YOUNG: Freeze water out every night.

SHOEMAKER: Stanley wants to know what happens to erosion if there is no running water but only ice. The answer is that this is an effective way to erode a planet; the erosion rates are high.

MILLER: Like Alaska, I think.

MURRAY: Freeze-thaw action helps erosion dramatically.

SHOEMAKER: Water can be moved around as snow, glaciers can be built up, resulting in very good erosion rates.

YOUNG: But there is not enough water on Mars—

MURRAY: We are trying to find out if a small ocean or permanent large lake in which organic compounds could have stayed in solution could be maintained over long periods of time. I suppose this could occur seasonally, but it might be a serious problem every year that it was frozen solid—all the way through. Chemical activity certainly would be inhibited. If what is needed is not necessarily a full ocean basin but the Caspian Sea—

MILLER: No, maybe 500 bodies of water, 10 miles across, the size of San Francisco Bay—

MURRAY: That is fine, but you might as well have oceans because the climatic conditions that support such open bodies of water have to be compatible with oceans.

HOROWITZ: The advantage of a terrestrial-sized ocean is its continuous existence for a long period of time, a long enough time to ensure an origin of life. Little puddles may not last long enough for anything to happen.

ORGEL: But if they happened every year, for a couple of weeks a year?

HOROWITZ: But on geological time scales it takes a couple of hundred million years to originate life; then these puddles are not going to be much good. Your reaction vessel must have a continuous existence.

RICH: As long as the reaction-vessel puddles are refilled periodically, why does it matter?

HOROWITZ: But the basin will not remain for a long enough time.

RICH: You mean it will fill up?

MURRAY: A basin must have enough history for a self-replicating molecule—or whatever else you define as the beginning of life—to get started. It must be there long enough to diversify enough to be able to adapt and survive the disappearance of the basin.

ORGEL: About 10-100 million years?

MURRAY: Yes, and that is not a seasonal melt on a glacier, that is a permanent body of water.

SHOEMAKER: If the melts occur at the same place every year it is all right.

OWEN: Could 100 million years be squeezed in here?

MURRAY: No. Water survives the annual cycle 10 or 100 or 1,000 years.

ORGEL: There is a terrestrial model. Some salty basins in the sea or someplace fill up once a year, stay full for 3 months, disappear again and then fill up in the following year. This occurs every year for 100 million years. That is Stan's point.

MURRAY: But when full, it is a free water surface that evaporates and pumps water into that atmosphere. On Mars such an open water basin won't form at all unless conditions are very different from what they are now.

OWEN: Right.

MURRAY: A greenhouse effect, and conditions necessary for it must be fulfilled. Central Australia is like this. It rains every 3 years and half the damned place gets flooded for 3 months and then it dries up. This occurs in places in Africa, too, sometimes, but water erosion is what is seen. You see water erosion and much moisture in those places.

MILLER: Would water erosion of that type be seen on the Mars' picture?

MURRAY: Yes. If permitted to persist 100 million years or even much less than that, water erosion would wipe out any topography not being renewed by uplift.

ORGEL: It could have occurred in the first 100 million.

MURRAY: Even if that situation were restricted to formation, moisture must have been in the soil. The earth is not completely drying out every 3 months. It finally gets wet enough for stationary water to form locally, but still there is moisture on the planet. Chemical weathering still goes on and when the storm comes, or whatever, it effectively serves to transport agents of weathering.

OWEN: You really have not discussed yet the problem of small-crater distribution and the evidence for some sort of erosional process on Mars.

MURRAY: This is irrelevant to what we are talking about now.

ORGEL: Why not squeeze in open water for the first 100 million years, then it is obliterated?

MURRAY: But the main point here is if the surface is primordial it *is* primordial and not primordial minus 100 million years.

ORGEL: How is that established?

MURRAY: I have already argued that the age of Mars cannot be 15 billion years.

ORGEL: Let's look at the logic. You calculate an age of 15 billion years based on assumptions, but you know the planet isn't more than 4.6 billion years. But do you really believe you can conclude from that that it must be 4.6 rather than 4.5? The model is clearly off by a factor of 3, but it easily could be off by a factor of 3.03.

MURRAY: But we are talking about physical events that either happened or didn't happen in the formation of these surfaces. Either enormous flux of objects hit Mars that didn't hit the moon in the last 4.4 billion years, and this flux simulates the event involved in the primordial formation of the planet, or it didn't.

SHOEMAKER: Bruce, let me try to reformulate and make the model compatible with what you are saying. Let Mars differentiate or have an atmosphere a la Stanley Miller; rocks melt the gas bubbles off early to make an atmosphere–which in the long life span of Mars is transient–that lasts about 100 million years. The surface is bombarded at a much higher rate than subsequently but on this surface there are scattered Orgel puddles.

ORGEL: Miller puddles.

SHOEMAKER: O.K., Miller puddles. Subsequently, as the atmosphere is lost, they dried up and disappeared—

MURRAY: Is this all during the planetesimal phase?

SHOEMAKER: Yes.

MURRAY: While 300-kilometer craters are forming, there is bombardment?

KAPLAN: Why should the atmosphere be lost at all if there was an original atmosphere? What processes would have caused the loss of an atmosphere on Mars?

OWEN: It could have been the solar wind in the absence of a magnetic field. But then, for reasons that are not very clear, it must be postulated that there was a magnetic field trapped in the early stages and subsequently lost as the planet heated up.

MURRAY: But we have argued against this: It would have modified the topography because of internal structure.

OWEN: No, the planet would not have to heat up enough to modify topography, that is the point.

MURRAY: Let me continue on my line of argument. What happened in between the 300 kilometer impacts I don't know. Probably it would be very difficult to maintain the conditions you need but I will ignore this.

OWEN: Why? It is important.

MURRAY: As the impacts occur the whole atmosphere is blown off.

SHOEMAKER: It might contribute to the dissipation.

MURRAY: If this is so, we should have been as excited about life on the moon as about life on Mars. Although the moon would not retain an atmosphere as long as Mars, it would retain it under those same conditions. Roughly the moon would have had the same kind of history as Mars.

OWEN: There is water on Mars now.

MURRAY: That water has gone.

LEOVY: Let me try another speculation in answer to Ian's question. Suppose with its present atmosphere the earth was moved out into the orbit of Mars. First all the oceans would freeze and the planet would have an albedo of 80 percent. After this, the water would rather rapidly condense out. In a very short time all of the water would effectively condense. The greenhouse effect would be reduced; further cooling, condensing, and trapping out of even the CO_2 would occur.

SHOEMAKER: That trap, the great pot of ice might be in Hellas.

MURRAY: Conway is giving a different reason for the absence of an earth-like atmosphere on the planet Mars: namely, it would be unstable at low insulation. I agree. Venus is probably stable the other way—too much insulation.

YOUNG: It might be a good idea to move earth out to the orbit of Mars.

KAPLAN: Yes, but this raises another problem. If differentiation is occurring now, it must come from internal heating. The heating must come from radioactive heat source and from what we know of mineral composition, potassium would be the most abundant element. If this is so we should expect to find argon.

Is there evidence for the presence of argon? Whether it has been produced over 3, 4, 5 billion years or not, argon must be present from a radioactive-breakdown process.

There is a very small magnetic field measured on Mars implying this differentiation—at least analogous to what has occurred on earth—has not happened. Otherwise, there would have been some kind of magnetic field.

SHOEMAKER: No. It is most likely that the circulation that produces the earth's magnetic field is from precession of the earth, driven by the moon. Without the moon it is not clear that the earth would have a magnetic field.

ORÓ: Bruce, could you please give us a more complete statement of your conclusion?

MURRAY: My first conclusion is that the present large-cratered surface is a remnant from the planetesimal phase of formation, similar to the interpretation of the craters of the lunar uplands. Second, there has not been an aqueous atmosphere, that is one in which stable bodies of water could survive, even over a 3-month period. There has not been an aqueous atmosphere since formation from planetesimals, although I cannot argue that there wasn't one before.

Third, there has not been a mobile crust like that on the earth since formation and, therefore, there has never been such a crust, in my opinion.

These are my conclusions. The probability is high that Mars is not an earth-like planet. Unless something happened in between the formation of the planetesimals themselves, it never had an earth-like atmosphere.

ORÓ: Therefore are the materials still inside?

MURRAY: Other than what is seen at present, it has not evolved an atmosphere yet.

ORÓ: But may it begin to now?

MURRAY: Yes, but this is speculation. The process of devolatilization may just be beginning now.

KAPLAN: Argon should be forming from potassium breakdown. Won't the measurement of argon tell us directly?

MURRAY: We know nothing about argon on Mars.

KAPLAN: Because it cannot be resolved?

MURRAY: Yes.

KAPLAN: It may be there as an important gas?

MURRAY: Yes. Toby Owen's prime wish with Viking was to measure argon. Measuring the total abundance of argon is probably the most important experiment we could do.

HOROWITZ: And nitrogen.

MURRAY: Argon is more important than nitrogen, because argon can't be fixed as nitrates. You can't fight that argument.

MILLER: You have assumed it takes a long period of time, 100 million years, to make life.

MURRAY: It takes a long time to evolve life compared to eroding unsupported topography.

MILLER: Could your topography be maintained with the liquid water for a million years? The million years might be enough to make life.

MURRAY: Probably yes,

MILLER: A year might be enough, but one year is a bit on the short side.

MURRAY: Why don't you accept the panspermia hypothesis and suggest Mars was contaminated by life from somewhere else?

MILLER: These are possibilities entirely within the framework of what we think took place on earth.

KLEIN: Much of your argument depends on the cratering on the moon. Yesterday I was confused as to where the moon was 4.5 billion years ago and what cataclysmic events accompanied the origin of the moon. Can you enlighten me?

MURRAY: Dr. Singer's highly speculative talk [pages 109-120] presumed the moon was in an orbit similar to that of the earth, so that the energy involved in capture was not very large so far as flux rates go.

KLEIN: Were there many little objects in this orbit which condensed into the moon?

MURRAY: I refuse to comment directly on yesterday's talk. But the moon differs from Mars in another way that may be very important. The moon has maria and the mass-cons that go with them. These features may not be represented on Mars. Certainly we see no evidence of Martian maria so far.

The origin of lunar maria is very hard to understand. One idea is tied into the double planet idea: a series of bodies, out at a certain distance. The moon went out to that distance early and got zapped pretty close by these few very large objects. Some kind of special creation must be invoked. Gene tried to avoid this by saying that the mare basin is formed over a very extended period by a random process.

The age of Tranquillitatis forces us to accept the fact that the mare basins were formed rather closely spaced in time. Therefore, they don't fall out of any random process. We are forced into supposing that maybe something in the earth-moon vicinity–for example, a ring of full-sized objects–went through, zapped it, and moved out a certain distance.

I mention this because it may be a further reason why the moon may have seen a much higher flux rate than Mars. Mars may not have had the double planet collection of debris around it to be bombarded by. If it was captured, it had to be captured from something near 180 orbit, so the flux rate arguments are the same.

MARGULIS: What is your argument for the difference in frequency of craters in Mars versus the moon? You say there are so many more.

MURRAY: No, the basic argument is that the extreme case states all the impacting obviously came from the asteroidal belt. Mars was preferentially hit by a factor of 20 to 25 to 1 which gives an upper limit. We know the flux rate on the moon but—

MARGULIS: Which hypothesis do you favor?

MURRAY: I think it is much too high because the flux that manifests itself in the present craters on the moon is not all asteroidal. There was considerable cometary impact, and contribution by comets may be related to a stable planet system such as is represented by the mare basins themselves.

Other objects that didn't get swept out that fast may have been around and contributed to some of the present craters in Mare Tranquillitatis. All this just drives the scaling down and, therefore, makes Mars older.

HOROWITZ: But the basic fact remains—if I may summarize the biological side—that there is an atmosphere on Mars. It is hard to produce this CO_2 atmosphere without producing several meters of water, as Toby argued this morning.

MURRAY: Ice.

HOROWITZ: Water in all phases. We can't decide if water was ever present as a liquid ocean. Water may never have been in liquid form on Mars.

MURRAY: That argument will work only if the CO_2 all formed back in the planetesimal stage when 300 kilometer asteroids were splashing among the puddles.

HOROWITZ: I don't agree. You cannot reconstruct the history of Mars back to the first few hundred million years.

MURRAY: We are discussing episodes. The episode forming the heavily cratered surface either could or could not have happened in the tangible history. It probably could not have happened. This not a gradational argument: There are certain probabilities for each specific history.

OWEN: But here is a good model for both atmosphere and crater production proceeding simultaneously.

MURRAY: The puddle-splashing model?

OWEN: Yes.

MURRAY: But if life can form that way, then it can form earlier. Why not form it out of the solar nebulae and deposit it by panspermia?

KAPLAN: If the big basin at Hellas which has no craters has been filled in now by sediment, sediment must be generated. Material cannot be moved that is not being generated. What process generates this sediment?

MURRAY: An even more profound argument for sediment and weathering is that the crater slopes on Mars are considerably smoother than on the moon. The Martian weathering process must be more effective than the moon's. It is a big mystery.

KAPLAN: This could be wind weathering, wind polishing by the dust itself.

MURRAY: Four kilometers of topography relief can't be broken down by 10 to 40 micron particles in the present Martian atmosphere. These won't wear it down. Something must be breaking up the relief. There must be a mechanism doing this which we don't understand.

KAPLAN: But it must be happening now.

MURRAY: Or it happens infrequently.

KAPLAN: It may happen relatively frequently to provide a source for sediment.

If the source for sediment is not wind, the only other known source that can generate sediments is liquid or fluid of some kind. What?

MURRAY: This is the problem. Chemical weathering soil moisture might explain a lot of things: breaking up of surface, low slopes, and so forth. The fact is that from what we know, water is not stable.

KAPLAN: Let me summarize the small crater history because it is relevant to this question. The small craters are much less abundant than expected on a pristine surface that had so many large craters. The secondary impacts alone from the large craters lead to an enormous number of small craters which are absent. No matter whether we are looking at something that happened recently—in the last few million years—or during formation billions of years ago, there is still a problem: There is a dearth of small craters no matter what the origin of the ancient large craters.

There are three possible histories which by themselves cannot be distinguished. At the end of the planetesimal phase there was a higher incidence of small particle bombardment—that sandblasted out 100-or-so meter craters. There is some evidence that this may have happened on the moon. This left the surface as it looks now. Since then the impact rates have been very low, slightly lower than on the moon. The erosion rates are trivial, so that the observed present small-crater abundance is somewhat less than Mare Tranquillitatis. The first possibility implies that the small-crater abundance measures all the impacts that have occurred since the end of the planetesimal phase. Net erosion, net burial, is trivial—because 4 billion years of the wind blowing around has not buried these 1-to 5-kilometer craters. This view also implies that the scaling is off and that Mars gets bombarded less frequently rather than more frequently than the moon does.

SHOEMAKER: This implies, too, a very different-size distribution of particles.

MURRAY: Not on the small craters, no?

SHOEMAKER: The problem is that the slope of craters in the size range of a few kilometers down to 300 meters, wherever the distribution cut off, is a dramatically different slope.

MURRAY: True, although we might invoke observational bias here.

ORÓ: No.

MURRAY: None of my three possibilities will work very well. This is important to consider because we don't really know that Mars gets bombarded more frequently than the moon, but this is the only way to exclude a mysterious erosion process. To avoid a mysterious erosion process you must accept this alternative one; if you don't like it, listen to alternatives 2 and 3. The second possibility involves again an early history of the cratered-planetesimal surface but a high erosion rate that wipes out craters as soon as they form. The erosion is discontinuous, it comes in bursts. The small craters are all removed every couple of million years or every 10 million years, so that persisting on the surface now are all the craters that formed since the last event 100 to 10 million years ago, and these do not show a distribution of erosion among themselves. That is one of the problems.

OWEN: What is the mechanism for bursts of erosion?

MURRAY: I have no idea. You tell me. I am pointing out a possible explanation of what is implied by the observations. This explains why eroded small craters aren't seen, which is one of the problems. There is no distribution of erosion among the small craters, they all appear relatively fresh.

SHOEMAKER: Because you can't recognize them.

MURRAY: I think with the B camera we could.

Alternative three involves a steady state, very strong erosional process that scrapes out small craters very fast, but does not destroy large craters. The time constant is 10-100 thousand years on a small scale. The small craters observed now are the result of recent activity, like a recent surge of asteroidal impacts. Either there has been a discontinuity in the recent impact history or in the recent erosion history, or nothing has happened. None of these is very attractive. Each requires something unusual on the planet. We already suspect something unusual with the erosion because of the chaotic terrain. We begin to be interested in the possibility that Mars has had a very episodic history.

Probably episodic vast erosion is the most likely possibility, although it is distasteful because we don't have real understanding of the mechanism.

OWEN: Could Hellas possibly represent an extreme example of whatever that process is?

MURRAY: Yes. Some mysterious process is going on with the small-crater history as well as with the chaotic terrain. For large areas of the planet apparently some peculiar kind of erosion is taking place—either discontinuous in time or continuous in time with a very high effectiveness with a discontinuous flux incidence.

OWEN: Only the discontinuous flux portion.

SHOEMAKER: But the crater distribution could be explained simply by some fairly steady erosion mechanism that is not, itself, connected with the crater.

MURRAY: If new Mariner '71 pictures show a distribution of types among the small craters that indicates a continuous population of formation and removal, then a very large continuous erosion rate can only be assumed.

We may be driven to admit this erosion occurs without understanding the underlying process. This probably represents the most important conclusion for biology. We see the fingerprint of something going on. The static model outlined here with CO_2 frost caps and primordial surface does not account for that erosional process.

ORÓ: Did someone suggest that the changes in the features that have been observed through the last 100 or so years bear some resemblance to the canals? A significant movement of dust particles on top of an intrinsic topography was more or less suggested. I'm not explaining this well but you should be familiar with the paper which interpreted the old fantasy in the light of recent observations.

MURRAY: They are old fantasies and are best left that way.

ORÓ: No.

SHELESNYAK: I am concerned about a local erosion problem; we have our own physiology to consider.

The conference recessed at 5:35 P.M.

SUNDAY EVENING SESSION
Dr. Bruce Murray, Chairman

GOALS OF FUTURE EXPLORATION

MURRAY: May we reconvene and extract from each one of us that last remaining erg of energy?

This evening we are supposed to imaginatively discuss the future—not just stick to this hard-headed, realistic discussion we have had so far. Is this what we want to do? Shall we consider how the '71 mission fits in?

RICH: Yes. What hard data information can we expect from the '71 that might solve some of the problems we have talked about?

MURRAY: I think the '71 mission will contribute to every point on our list. The chance to map the variation of the large and small craters across the planet, to see to what extent the surfaces are of different ages, to understand the boundary of Hellas, to see in detail what happens at higher resolutions at the cratered-uncratered terrain discontinuity, to find the signature of the processes there. Mariner '71 should produce more than 100 times the total data, up anywhere from a factor of 2 to 7 greater resolution. It will contribute very greatly to this question of the history and the processes which may be involved.

HOROWITZ: Seventy percent of the planet will be seen at these resolutions and practically the entire planet will be mapped.

MURRAY: This mission is specifically intended to look for clouds and, if found, to use some time to monitor the clouds' diurnal behavior at reasonable resolution, the so-called cloud-chasing mode.

MILLER: Can it determine the water concentration as a function of latitude?

MURRAY: Yes. There is a microcentric interferometer experiment from Nimbus which is very sensitive for the 6-micron water band. It will attempt to map the water vapor distribution around the planet on a daily basis.

YOUNG: It is very poor resolution.

MURRAY: Yes, the spatial resolution is several hundred kilometers. Microcentric interferometers flown remotely are very complicated and uncertain, but this one, based on a Nimbus system, hopefully will work all right.

OWEN: What is the sensitivity of the spectrum?

HOROWITZ: It is better than the Mariner '69 IR spectrum.

MURRAY: Yes, by 1 percent. That had a resolution of 10^{-2}, not exactly something to write home about. I am sorry, I don't know, but the '71 is higher. It is certainly high for the water-band question.

OWEN: It is high enough to really go to work?

MURRAY: Oh, yes, it is optimized and justified.

HOROWITZ: I though it was primarily a temperature probing.

MURRAY: No, primarily 6 micron water.

HOROWITZ: I am glad to hear that.

LEOVY: In the Nimbus experiment the microcentric interferometer does serve primarily as a temperature-broken CO_2 band.

MURRAY: Perhaps I am wrong but the reason it was chosen for this mission was its water vapor, not its temperature capability.

YOUNG: Yes, it had a better capability for water detection than other proposed spectrometers, but it really wasn't that good. Although modifications have been made, I think it still isn't that good for water.

MURRAY: An infrared radiometer is included which measures surface temperature distribution into the nighttime side allows us to make a complete thermal map of the surface.

To measure nitrogen, we have the same experiment we repeated many times which probably doesn't help.

MILLER: It is no more sensitive?

MURRAY: It is the same instrument.

YOUNG: Wasn't this instrument modified?

MURRAY: Yes, but it is exactly the same experiment repeated on the limb. How many times is it worth repeating because something must be given up when the limb experiment is done? There is a horror story associated with the slit alignment of that instrument and the others on the platform with it.

SHOEMAKER: What is the limb experiment?

MURRAY: The ultraviolet spectrometer. It will improve the lower limit for nitrogen somewhat.

HOROWITZ: It operated so well that it would be hard to improve on the '69 results, so it has nearly worked itself out of a job for '71.

OWEN: Then how will it help?

MURRAY: It is closer to the planet so it may be more sensitive. The scan platform may be cocked away to get a better pathway. The mission is relatively adaptive so that results in the early portion can influence experiments in the later portion.

HOROWITZ: Charlie Barth probably has a million things he wants to do.

MURRAY: The temperature mapping may help the carbon cycle question somewhat but not terribly much. Frankly, unless there is some transitory variation in the CO_2 or CO this probably will not contribute beyond what is known.

MILLER: Will it map the height of the CO_2 path?

MURRAY: Pardon an impious remark, but the only useful aspect of the infrared spectrometer aboard Mariner '67 *was* that although it was not intended. It is not aboard Mariner '71, and I don't know if the infrared interferometer will map that.

In principle the ultraviolet device can get it also. It can repeatedly separate radii from surface information but it is kind of weak.

In terms of height, the 50 or so occultations at various latitudes and longitudes will surely provide some nice geometric control on the height question.

OWEN: It is very unlikely that any low areas will be seen.

MURRAY: Of course. The highest ridge between the point of occultation and the earth will be seen when it happens, but over such a range that you should get pretty good control on it.

OWEN: But we won't know what we want to know—how long it goes.

MURRAY: Combined with the radar, we're O.K. We already have some pretty good data from the ground-based radar.

On the light-and-dark-area question, the mission is supposed to spend great effort studying seasonal variations, but I'm not as hopeful as I was. With photometry, they will again monitor the same area at different times with the same lighting. Careful experiments are planned, but this is very difficult. These areas will be viewed for the presence of and then the variation in water vapor so if anything like those rimmed craters are in the light and dark areas or associated with the fading of the dark areas, this should be detected.

HOROWITZ: The main trouble is, it comes so late that the darkening won't be seen. It will see the fading in most places.

MURRAY: This is difficult.

Gravitational information may help us with the internal history question. Large free air anomalies supported by the planet should be detectable. Once the whole planet has been mapped we can determine whether or not there are anomalous tectonic features.

MARGULIS: Do you have any specific plans to look at the erosion problem?

MURRAY: We hope to study the erosion as a function of latitudes at the very least.

MARGULIS: By taking pictures?

MURRAY: By interfering slope information. We've had difficulties with the '69 data in the reconstruction of accurate photometry across the pictures. We are hopeful that good photometry will come out of the '71 pictures; then with the slope information we may see latitudinal differences in erosion.

MARGULIS: What is the best possible result you might get from slope information?

MURRAY: This relates to understanding the weathering process. The weathering makes the slopes smaller and if we could find differences with latitude of, say, temperature, it might give us a clue. If the slope spectrum is the same all over the planet, water is probably not involved because water weathering should operate differently in the polar areas than in the equatorial. If we find no significant latitudinal differences—

MARGULIS: You might be able to eliminate water.

MURRAY: Yes, or we might be able to find evidence for it.

RICH: Are the science packages on both orbiters identical?

MURRAY: Yes, there are two different orbits; it is not clear which spacecraft orbits.

If temporal variations of surface radiation occur, the UV experiment repeated over and over will provide information on that.

If any transitory atmospheric or surface features are observed, they will be followed. This is intended for exactly that purpose. This is true also for anomalous frost and clouds.

The oxygen question will not be affected by anything that I can see. The question of surface composition won't either, although there is armwaving about rescron—

RICH: What?

MURRAY: Rescron. Anomalous emission or anomalous reflection in the infrared arising from stretch vibrations of the silicon-oxygen bonds and silicates–is called rescron. "Residual ray bands."

Fifty samples of the occultation, although limited by the highest occulting topographic features, nevertheless will certainly provide measurements of surface pressure with elevation.

Time and space variations can be studied because of the 90-day lifetime. Presumably some occultations will look at the same general area at different times.

The significant point is the amount of returned data that is expected in the '71: a factor of a hundred or more total data returned spread over a factor of a hundred in time. We expect great things.

RICH: How many orbital changes will it be capable of?

MURRAY: Four or five, limited by the number of valves that can be turned on and off. Very large orbital changes cannot be made. Sterilization is relevant: The flexibility is determined by how close we will be to the planet.

HOROWITZ: But there is time to change that quarantine requirement.

MURRAY: Not really, not unless it is done pretty soon.

RICH: Why?

MURRAY: Because they are looking into the mission profile. The project program manager at NASA headquarters said he would not support any activity by the JPL project to investigate the consequences of lowering that periaxis height below that presently believed to be consistent with quarantine.

RICH: Because he already has his program, and he doesn't want to rewrite it?

MURRAY: He doesn't want to rock the boat. The quarantine is a sacred cow within NASA.

YOUNG: This was reviewed by the Space Science Board and they recommended no change.

MURRAY: There was no presentation made of the consequences of leaving the plan the way it is—because the work hadn't even been done by then.

STROMINGER: Is the fear of contaminating Mars?

MURRAY: It is based on keeping it in orbit until 1985.

SHOEMAKER: The justification for the manned mission to Mars will be to go scoop the orbiter up.

MURRAY: No, the justification is that if we have manned missions by 1985, we don't have to worry about keeping it clean.

RICH: Beer cans will be on Mars by then.

HOROWITZ: Biologically, maybe one of the most important things this orbiter will do will be to find out if these brightenings Conway Leovy discussed are clouds.

LEOVY: It should resolve the question of clouds any distance above the surface.

HOROWITZ: It will pinpoint possible sites for an eventual lander.

STROMINGER: If you had to choose now, at what site would you land?

HOROWITZ: In the Tharsus Candor region where all these light streaks are.

SHOEMAKER: Why?

HOROWITZ: Because by Martian standards they could be wet places.

KLEIN: The bright streaky areas, not what Conway was talking about but the spaghetti-like stuff?

HOROWITZ: Yes, but apparently it is all part of the same system.

KLEIN: Could it be related in any way to the Hellas type of terrain?

MURRAY: No. There is not any similarity except that both are unexplained and both are brightenings.

HOROWITZ: The Hellas brightenings and the Tharsus Candor brightenings are both diurnal, appearing at the same time of day.

KLEIN: In the afternoon?

HOROWITZ: Yes.

OWEN: But the Hellas brightening doesn't occur daily, sometimes it is there and sometimes not.

MARGULIS: Bruce, if the quarantine people allowed it, how much closer to the planet would you want to go?

MURRAY: I can't answer because they won't even let the question be analyzed.

MARGULIS: What must be weighed against what in the decision not to get closer?

MURRAY: Around 400 kilometers there is a gain of resolution by the system, closer you begin to get smear limit enough to question the desirability of going lower.

HOROWITZ: Resolution is improved by a factor of 2.

MARGULIS: Is the limit 400?

MURRAY: If quarantine were no worry and we had an aggressive policy to explore the planet, the biggest step would be going in at around 400 kilometers.

MARGULIS: Is quarantine preventing this?

MURRAY: We think so. There could be other factors: Thermal control, solar occultation, but we can't even get this analyzed.

RICH: Have you power to go up to 800 if you go into 400?

MURRAY: Yes, that is easy; the reverse is hard. The entry point at the moment is 1150 kilometers because–again, a sterilization question–this gives a 10^{-4} probability of impact.

This all has to do with P_g, the probability of growth. What is the probability of growth of a terrestrial organism? Assuming it is unity, or 0.1, implies a very conservative engineering strategy. If ionizing radiation makes the probability of organisms surviving and growing much lower, then it is easy; no problem. An order of magnitude change would make the problem go away.

YOUNG: A new look will be taken at the P_g question but I doubt if it will be in time for this mission.

MURRAY: Yes, because nobody is pushing it. Frankly, it is a political question: The NASA administration had made it clear that under no circumstances will they initiate any action in this regard. They refuse to be accused of dirtying up Mars even if it means doing a less-good job scientifically. This is their strategy. The Space Science Board is acting inept and unresponsive and it feels that it has no responsibility in this area and so goes along with NASA.

MARGULIS: What if you have a very impressive petition with a long list of signatures and whatever else it takes politically to get down to 400 kilometers?

HOROWITZ: Murray and Horowitz are the only names.

RICH: Unfortunately that is not the way things get done.

MARGULIS: What do you want, agitation—a riot?

MURRAY: This is an illustration that the scientific advisory structure of NASA is dominated by non-participants. This happens when weekend warriors without a stake in the mission are on the Space Science Board. They go along with peculiar restrictions that they would never put up with if they were really involved.

RICH: Does the LPMB—Lunar Planning Mission Board—have any voice in this problem?

MURRAY: If you were doing your job you would have raised the question a long time ago.

SHOEMAKER: We ought at least to be able to holler about it and find out how much voice they have.

MURRAY: You might also ask why there is only a single launch for the one flight to Mercury.

RICH: One problem at a time.

MURRAY: There are many problems like this with programs other than Viking that the LPMB hasn't looked at adequately, decided what the issues are, and really forced them forward. The rest of us are quite helpless on most of these.

Enough on the quarantine. The quarantine is an inhibition both in a direct sense and, more importantly, an inhibition psychologically because there is strong antipathy to even raising the issue.

RICH: If the orbiter enters at 1,100, the chance of dropping down to 400 and getting up again to 800 successfully is good.

MURRAY: But they will say that we must go down and get up with 10^{-4}, that is, a P_g of 0.999.

RICH: There are two ways of thinking about the problem. What is the likelihood of working now before you are inserted and once you are inserted, what is the likelihood of being able to go ahead?

MURRAY: But since it has to be a .999 probability that all these maneuvers will succeed serially, I think there will be difficulty getting an answer. The problem with sterilization is that it is based on unrealizable objectives and nothing can be done. An absurdly conservative strategy must always be followed.

RICH: This isn't primarily sterilization, but more a question of the reliability of the instrument once it is operating.

MURRAY: It is very model-dependent, too; it depends on the density of the atmosphere.

RICH: What is the difference in arrival time between the two?

MURRAY: About a week, I think.

RICH: Would information from the first week in orbit at 800 kilometers be enough?

MURRAY: It can't be known until you are there. This is irrelevant, angels dancing on the head of a pin. The problem is P_g—the probability of growth. If you believe the probability of growth is unity or tangible for a terrestrial microorganism on the surface of Mars, this is a reasonable strategy. You might believe that the evidence indicates estimates for the probability of growth are very low, that one of those anaerobic, nitrogen-hating, UV loving—all these adjectives biologists like to throw around in series—organisms is going to get down. These are not even the kinds of bugs on the spacecraft in the first place. If you estimate high probability that this happens in series, and the perfect bug finds a little niche and spreads and contaminates the goddamned planet in 20 years—then this policy is not bad. But if you believe this is an extraordinarily extreme version, and that realities are much less—

MARGULIS: P_g equals zero.

MURRAY: Or even 10^{-2}—the problem disappears. The engineering problems become dominant. I claim the issue ought to be thought out.

RICH: I understand.

MARGULIS: Who besides Alex is in a position to do something about this?

RICH: Gene Shoemaker.

MURRAY: The biologists themselves originated this problem and it is very difficult for a nonbiologist to change it. This must come from people—

RICH: I spent some most discouraging days in Washington with a subcommittee. Were you there?

YOUNG: Which one?

RICH: We spent a whole day going through a P_g calculation.

YOUNG: It will be done again.

RICH: It was a real fantasy.

MURRAY: The Greeks had a capital offense called sophistry and although in the case of Socrates they killed the wrong guy for it, I believe it really is a capital offense.

STROMINGER: My opinion is that P_g is extremely close to zero and I would fly as close as I damned well wanted to—on the theory that it is possible that this may turn out to be our last chance to look at Mars. I would get every bit of information I could.

MILLER: To take a risk and say P_g is really going too far.

MURRAY: Yes, this isn't necessary; it can just be 10^{-2}. That is why we are asking what it will cost the project to find out.

HOROWITZ: We ought to correct it, the actual P_g is more like 10^{-2} or 10^{-3}

MARGULIS: "g" is meant in the sense of exponential growth rather than survival, isn't it?

HOROWITZ: Yes, growth.

STROMINGER: It is somewhere between 1.0 and 10 to the minus infinity and, I don't see how it can be estimated.

YOUNG: Estimates can be made, experiments have been done simulating the environment of Mars to see how terrestrial organisms behave.

MURRAY: Wait now, this P_g was arrived at years ago and it hasn't been changed since the COSPAR regulations of 1964.

YOUNG: For the end product, 10^{-3} probability, that number has changed.

MURRAY: On landers only, it was relaxed an order of magnitude. Only 1/1,000 of one bug per lander is allowed—changed from 1/10,000 of one bug per lander—but the probability of contamination by a fly-by or orbiter remains the same as it was formulated in a rather poorly thoughtout session in 1964.

HOROWITZ: It is an undocumented number.

OWEN: Who is responsible?

MURRAY: Carl Sagan from Cornell, Joshua Lederberg from Stanford.

HOROWITZ: Allen Brown.

KLEIN: Although Brown was not a co-sponsor, he certainly is one of the people behind this argument and its philosophy.

MURRAY: Yes. I have argued with him, told him he is making it likely that the first events will be Soviet cones and since the Soviets don't sterilize to near what we consider adequate standards–even by my standards–these curious attitudes may inadvertently lead to contamination. He acknowledges this reality but somehow this doesn't seem to change his conclusion.

RICH: What kind of sterilization procedure is the orbiter going through?

MURRAY: None. There is no decontamination procedure.

YOUNG: That is why they can't lower the orbit.

RICH: Aren't they even spraying with ethyl oxide?

MURRAY: But it is sitting out there soaking in a vacuum for 6 to 9 months, 27 years, and even if they sprayed, it wouldn't work because organisms will be buried in the potting of resistors and things, and maybe these will be broken open anderoded and release the bug.

STROMINGER: What is the probability of crashing?

HOROWITZ: That is all figured in.

SHOEMAKER: About 100 percent.

STROMINGER: The probability of crashing is so much greater.

OWEN: That is the whole point of the constraint on the orbit.

MURRAY: They aim away from the planet. When they aim for the earth, they aim 100,000 kilometers away from Mars and then shift midcourse moving a little closer. These guys take this seriously.

RICH: Once safely inserted in orbit up at 800 after having already fired things and everything is working: At that stage what is the probability of descending to 400 kilometers and safely coming back up again?

MURRAY: That's taken into account. Based on our present knowledge of the Mars atmosphere, there is no possible way to have adaptive approaches with this kind of launching.

RICH: Can't something be learned after a week in orbit at 800 kilometers?

MURRAY: We are talking about decay in 27 years. That is a fantastically small deceleration. The P_g requirement is a probability of 10^{-4}, but not until 1985. It is angels dancing on the head of a pin: terribly out of proportion with the cost, difficulty, and need for scientific information to know where we are landing.

KLEIN: At the 1,200 kilometer distance, how will your resolution in '71 compare to that in '69?

HOROWITZ: Previously we were at 3,000 and 1,600 kilometers.

MURRAY: It would have improved by a factor of 2 better. 800 kilometers–which is not even guaranteed–would improve it by a factor of 4 better, but we could go down 400 to make resolution a factor of 8 better.

MARGULIS: Is that as far as you want to go?

MURRAY: There is a smear limit as well as other factors. At these heights sterilization really isn't involved anymore.

KLEIN: Will the spatial resolution at 800 be about 60 meters?

HOROWITZ: A little less than 100 meters.

MURRAY: The low contrast target is not with shadowing, the resolution is a little better than half a kilometer. Going down by a factor of 4 gives us 100, 125 meters; by a factor of 8, we resolve about 60 meters.

MARGULIS: Can this be attacked as a political decision? Have you a list of people who have to be lobbied?

MURRAY: I have wasted enough time on this subject; I have written articles in *Science*. The decision is clearly stupid. I conclude this is a type of irrationality comparable to astrology and many other aspects of modern society that I don't understand.

KLEIN: Do you think the orbiter will be fired at the right astrological moment?

MURRAY: Maybe Ronald Reagan will be President and will appoint an astrologer to tell us when to fire.

KLEIN: Last week Conway Snyder at JPL discussed a new film system that may be used on the Venus-Mercury thing. If it is used in the orbiter of the Viking '75 mission it would apparently resolve about 15 meters. It is too late to adapt that type of instrument to the '71?

MURRAY: Yes. Much too late for the '71, but maybe not too late to adapt it to Viking. Here is a different statement, a precept. I am writing a book on the subject. On a photographic exploration, limits are set first by the total information returned. That determines the maximum resolution. At each new slug of information the attempt should be to make a big step in resolution. '71 was not pursued this way; it has been very docile. We have pushed from 1600 to 1800 and maybe that is it. It also has been clear for a long time that with Viking, too, the resolution is unfavorable. Stupidity again. $200 million is spent for an orbiter, at least some technologically possible decent photography should be given to the capsule site. Again this is part of the complex decision-making structure. The obvious first order problem was not embraced. In the case of '71, the quarantine has and still does contribute to that problem.

I don't know what the problem with Viking is and why people have not more aggressively done a better job.

Are there other comments on the '71 mission that are not likely to turn me on like this one?

MILLER: What is the total cost?

MURRAY: $150 million: $100 million for the spacecraft, 2 Atlas Centaurs at about $36 million, tracking data acquisition about $150. Mariner 6 and 7 were about $50 million; Mariner 4 was $120 million. Allowing for inflation it costs $150 million to send a remote sensing mission to Mars, whether or not you orbit. In terms of photographic data the cost per bit is about the same; there were two orders of magnitude increase between Mariner '69 and Mariner 4. In other words the price goes down by two orders of magnitude per shot; the cost per bit for photographic or other data on Mariner '71 is 10^{-4} what it was on Mariner 4.

RICH: If plugged into the GNP it looks even better.

MILLER: Will the trend continue to the Viking?

MURRAY: No. Viking doesn't get any more data and the cost goes up, so this trend reverses if they get the data from the pyrolysis gas chromatograph-mass spectrometry.

KLEIN: But this is qualitatively a different kind of data. Perhaps you want to go down to the surface instead of 400 kilometers?

MURRAY: No, surface imaging is different. The Surveyor versus the lunar orbiter is a good example of that. The Surveyor showed only a very limited number of things, although some exquisitely well, whereas the orbiter is really the most powerful exploration tool from a photographic viewpoint.

HOROWITZ: But to measure the atmosphere it is of course important to land on the surface.

MURRAY: Yes, if you land on the surface, but in '71 this is still a remote sensor question, and there is no substitute for good resolution.

SHOEMAKER: Must coverage be given up if the orbiter descends to 400 kilometers?

MURRAY: Some redundant coverage; some of the side lap is given up.

SHOEMAKER: But could 70 percent of the planet still be covered?

MURRAY: Yes, pretty close to 70 percent.

ORGEL: What are the limits on going down progressively? Say at least 100 miles from 800?

MURRAY: It is a numbers game. There are no new observations because of such small decay rates.

ORGEL: Even if you can stay in some orbit for a while, can't you still show that you can get another 100 miles down?

MURRAY: But we have to stay in that orbit until 1985 or 1987, or whatever.

SHOEMAKER: Twenty-seven years means until 1998.

MURRAY: Some long time, it doesn't make much difference.

KLEIN: Why so long?

MURRAY: So the damned terrestrial bugs won't get down where they can grow and contaminate the planet before we carry out life exercises. The logic is all related to P_g–the probability of bugs growing.

HOROWITZ: In spite of the fact that there are places on earth, paradise compared to Mars, where bugs don't grow.

ORGEL: Have simulation experiments been done?

HOROWITZ: None of the simulation experiments have been very realistic. Too much water is always added.

MURRAY: The dry valleys of Antarctica–a natural desert on earth–are essentially abiotic and soils there are sterile; they contain no micro-organisms. Nothing at all can be cultured from them.

ORGEL: Wouldn't it be persuasive, if unscientific, to smash up a spacecraft in the Antarctic and show that nothing happens?

HOROWITZ: Those places are continuously being infected.

ORGEL: Norman, you are being rational which is not good in this situation.

HOROWITZ: I have discovered that already.

ORGEL: Suppose we smash up, drop 100 or 1,000 organisms and show that none divide–doesn't that answer something?

MARGULIS: It is being done daily.

HOROWITZ: People set up camps, leave all sorts of garbage, and nothing grows.

KLEIN: Wouldn't a GCMS pick up the garbage?

MURRAY: We are not proposing that we drop garbage on Mars.

KLEIN: But a spacecraft dropped on Mars is not just bugs.

HOROWITZ: That spacecraft would never be found again.

OWEN: That is not the problem.

MURRAY: The problem, the bugaboo, is growth. I am afraid my ability to take seriously the arguments of biologists about life on Mars has been influenced over the years by this particular ridiculous set of arguments. I have had difficulties separating those two reactions and may have been overbiased in that regard.

YOUNG: Are you implying the probability of terrestrial organisms growing on Mars is zero?

MURRAY: No, but it is very, very low.

MARGULIS: Growth and survival are different.

HOROWITZ: What is relevant is the probability of accumulating sufficient mutations in one cell to permit it to grow in the Martian environment. I doubt if any organisms among those commonly occurring on earth could grow on Mars. It would take new mutations–I would estimate the probability of the order of 1 in 10^{16}.

YOUNG: Assuming there is no water.

HOROWITZ: Assuming only that Mars is less favorable environment for terrestrial microorganisms than the Antarctic desert, which is within a few miles of the ocean, and has glaciers and water and permafrost within 12 inches of the surface.

YOUNG: But you don't know anything about Mars. I am not arguing P_g is 1. I never have. As you know I lean more your way than most who have been involved, but I certainly don't lean all the way to zero.

HOROWITZ: I'm not saying zero, say 10^{-10}.

MURRAY: But even worse: what time scale for the P_g? Suppose a bug or two could grow, how long would it take to—

OWEN: To cover the planet?

MURRAY: Yes, to infect the natural Martian life?

YOUNG: I don't know.

MILLER: That could be very fast.

MURRAY: It could be, but with what probability? All the serial probabilities have a much smaller number than the one used.

MILLER: Once it gets going it simply propagates exponentially with the division time.

MURRAY: For example, the lithium chloride brine solutions are there, but it will take a while to blow around into lithium chloride. So it contaminates that and one brine is filled up with *Bacillus coli,* or whatever, how long does it take to spread over the whole goddamned planet?

HOROWITZ: Bruce, you lose your case if you admit that: You had better keep quiet.

MILLER: Yes, you are much better off with the P_g argument.

MURRAY: The implication is that these factors are unity and the others are not. Any reasonable integration over the ensemble possibilities leads to a very small number.

HOROWITZ: You should consult counsel before you make statements.

MURRAY: I have given up. I conclude this is a kind of irrationality that pervades the society we live in.

YOUNG: It is not irrational; it is simply overdone.

MURRAY: That's why it is irrational. When you tie your hand behind your back, it is like putting a lander earlier, which was irrational.

YOUNG: Can one of your biologists be put on the Space Science Board?

SHELESNYAK: It is remarkable that we are so much more concerned, fortunately, about contamination of Mars than we are about earth.

MURRAY: In a more serious vein, this is an illustration of a weakness in our capacity as an intellectual group to participate in society.

SHELESNYAK: You are pursuing the basic myth that man is a rational organism.

MURRAY: This is different, and particularly offensive because it is a case of a self-imposed restriction by scientists on scientists.

SHELESNYAK: But scientists are men and not rational organisms. The evidence, if it exists at all, is relatively limited that we are rational.

HOROWITZ: We haven't much time left and you promised to give us your pitch for why Mercury is a better place for life than Mars.

MURRAY: I think the probability of life on Mercury should be the same order as that on Mars.

YOUNG: No, you said you *proved* that Mercury was a more likely place—

MURRAY: This argues two ways. To the extent that the early existence of oceans or similar stationary bodies of water is important, I argue that the chance of that happening on Mercury, based on what we now know is greater than that it would have happened on Mars. The evidence tends to exclude open water episodes in Martian history, although it does not bring the probability to zero.

We feel that on Mercury there must have been some event involving chemical differentiation either in a nebula before Mercury was formed or shortly after it was formed. This was needed in order to end up with the analysis of high density we now see. On the basis of present ignorance there may well have been an episode of events comparable to what Stanley was talking about where gaseous-bearing material is brought in, heated up, gives off an atmosphere, and so forth.

If a mission is ever flow to Mercury I think it very reasonable to look for remnants of an early history in terms of mountains or something, but I would be astonished now to see mountain belts on Mars.

RICH: What is the surface temperature of Mercury?

MURRAY: The present surface temperature is a maximum of about 600°, but it is much lower in polar areas.

MILLER: Like what?

MURRAY: Down on a level surface it may be perhaps 200° Kelvin.

RICH: What is the rotation period of Mercury?

HOROWITZ: Once per year, it keeps one face to the sun.

MARGULIS: Not anymore.

OWEN: It used to but it doesn't anymore.

MURRAY: Anyway, this is only a semi-serious argument.

RICH: What is the circumference of Mercury?

MURRAY: It is 5000 kilometers in diameter but it is dense, its gravity is somewhat higher.

YOUNG: It essentially has no atmosphere.

MURRAY: Yes, I think that is true, but we really don't know. The upper limit is about a millibar. I think the chances are probably higher that there were oceans on Mercury than on Mars. The chances are at least comparable on Mercury, and higher on Venus. Venus may have had an ocean at an earlier stage.

The limitation to organisms on Mars involves the temperature and lack of water. The subsurface of Mercury, unlike the moon or Mars, is always above freezing over the large equatorial areas at all depths. It never gets down to freezing at all, except in the polar areas. Consequently any water that was liberated from the interior—by whatever process—could have reached the surface or could have been retained if the permeability varied as soil moisture.

The availability of liquid water in the soil, although not at the depth to which sunlight penetrates, is *a priori* much better on Mercury than on Mars. I don't think there is life on Mercury but I am trying to show how thin I think the argument is for life on Mars by comparison.

RICH: Someday you may see a well-formulated and carefully developed scenario supporting the idea of life on Mercury.

MURRAY: If it had a millibar atmosphere, then it would be a far better target. At the time the Space Science Board picked Mars as a target it was thought to have a millibar atmosphere, further evidence of the thinness of the view.

RICH: This argument will be used to support the idea of a manned program to Mercury.

MURRAY: I think the six suns one sees going to Mercury tend to keep it a pretty hostile place. I am serious that either way it is a long shot. We know so little about Mercury and if it turned out to have a millibar or two of neutral atmosphere, it might be a candidate.

Some people think it has an atmosphere, in that case it suddenly would become a much better bet. The probability that Mercury has a millibar atmosphere certainly cannot be less than 10 or 20 percent. If it has, it automatically is a much better place to look for life than Mars.

MILLER: Let's grant a millibar or 10 millibars atmosphere–still the temperature gets to 600°.

OWEN: At the subsolar point at the equator but not the complete spectrum.

MURRAY: What temperature do you want?

MILLER: 37°C.

SHOEMAKER: There are extremely cold areas on Mercury.

RICH: At 25° north latitude you can oscillate between 37° in the daytime when facing the sun.

MURRAY: In a crater near the polar area it is completely shaded.

RICH: But it goes around behind.

MURRAY: Near the polar area steep-sloped craters are permanently shaded on the inside; they are ice chests.

SHOEMAKER: If you go to high latitude, you can pick your maximum temperature.

MURRAY: Mercury can provide me with any environment I want.

OWEN: The strongest argument against the heat hypothesis is Mercury's high density. You will have a hard time driving out volatiles from the solid body.

MURRAY: But how much volatile material escaped from the cloud beforehand and how much afterwards? State your argument so that Gene can hear it.

OWEN: Bruce claims Mercury's high density is an indication of differentiation of material and, therefore, maybe—

SHOEMAKER: Differentiation of the material would have had to occur before condensation or else be blown all away.

MURRAY: Or some mixture of both, just like the argument for Mars.

OWEN: How is material really blown away in a way that really changes the body density?

SHOEMAKER: It is pretty hard.

OWEN: The probability is high that it occurs before formation.

SHOEMAKER: That doesn't mean Mercury is undifferentiated.

OWEN: No. It means this spongy mass of water under the surface is less likely. If volatile loss occurred to result in a high density planet, water vapor, too, would have blown off.

MURRAY: We know the surface material, at least the outermost portion, of Mercury is silicates because of the optical and the radio properties. Mercury may be like the earth with a much bigger core, and lacking a dense atmosphere. Given the odds and our ignorance it certainly is a possibility. This possibility is more attractive than the planetesimal-pole hypothesis we discussed.

OWEN: I don't agree. In both cases we don't know, and are just speculating.

MURRAY: The range of possibilities with Mars is being pushed back earlier and earlier. It is getting more and more difficult. Mercury might have been very earth-like early in its history, remnants of mountain belts might still be there.

HOROWITZ: Life couldn't exist on a planet with no atmosphere.

MURRAY: We don't know that it didn't have and doesn't now have an atmosphere.

HOROWITZ: I am talking about now. Life must live on the surface of a planet, see the sun.

MURRAY: If Mars ever were earth-like, it must have been in an extremely brief, very restricted period of time. This we know for Mars but not for Mercury. They are both long shots. I am facing this cultural lag problem: If Mercury was looked at as carefully as Mars was in '62 or '63, Mercury was at least as good a bet as Mars was for harboring life.

HOROWITZ: But Mars had an 85 mb atmosphere.

MURRAY: Yes, Mars had 85 mb atmosphere, water ice, polar caps, which moved seasonally, seasonal changes thought to be due to plants were observed, and it had Sinton bands—evidence for organic compounds—and all this has disappeared. Enough said. I have made my point.

RICH: It had more than that: Martians were well established in everyone's psyche. But now you are preaching to the converted.

SHELESNYAK: Mars had a good transportation system with canals and super-highways.

MURRAY: I remember when the pictures from Mariner 4 came in, the press asked "What about the canals, Doctor?" At the time I hadn't read up enough Lowell to appreciate why there was such a strong interest in canals. There had been this long cultural gap that I didn't share.

SHELESNYAK: Before we end tonight, I want to thank everyone for participating. We are not sure about when our next session will be or even if the subject will be biogenesis. We will let you know.

I want to thank you all for your endurance. We don't usually work 16 hours a day but this shows a tremendous capacity for biological survival under adverse conditions and it perhaps should encourage us in the thought that some organism might be able to survive on Mars and Venus.

HOROWITZ: I thank the two discussion initiators Gene Shoemaker and Bruce Murray, and everyone else who has participated at my invitation. Bruce, you probably did make an impression here, even on those who don't accept your whole thesis. Eventually your views will have an effect on our thinking about how the Mars problem ought to be approached.

I don't agree with you completely. It is worth going to the surface of Mars to find out whether life is there. I believe your arguments are weighty. And no one doubts that the *a priori* probability of finding life on Mars has dropped as a result of the data. You think it has dropped almost to zero. I don't, but I agree it has dropped.

It is a value judgment–how much time and money you are willing to spend to look for life on Mars. For the biologist, this is an extremely important question. The Viking Project would cost about one week of the Vietnam War. If you think of it that way it is a drop in the bucket.

MURRAY: But the other criterion is how many weeks of the NSF budget Viking would cost.

HOROWITZ: You mean years: It would be about 2 years of the NSF budget.

MARGULIS: On that lovely note of our national priorities, we can close.

The conference adjourned at 9:15 p.m.

REFERENCES

Adams, J. B. 1968. Lunar and Martian surfaces: Petrologic significance of absorption bands in the near-infrared. *Science* 159:1453.

Alfven, H. 1965. Origin of the moon. *Science* 148:476-477.

Anders, E. and Arnold, J. R. 1965. Age of craters on Mars. *Science* 149:1494-1496.

Barth, C. A. 1969. Planetary ultraviolet spectroscopy. *Applied Optics* 8:1259-1304.

Barth, C. A.; Fastie, W. G.; Hord, C. W.; Pearce, J. B.; Kelly, K. K.; Stewart, A. I.; Thomas, G. E.; Anderson, G. P.; and Raper, O. F. 1969. Mariner 6: Ultraviolet spectrum of Mars' upper atmosphere. *Science* 165:1004-1005.

Cameron, A. G. W. 1970. Formation of the earth-moon system. *EOS* 51:628. (*Transactions of American Geophysical Union*)

Cutts, J. A.; Soderblom, L. A.; Sharp, R. P.; Smith, B. A.; and Murray, B. C. 1971. The surface of Mars 3. Light and dark markings. *J. Geophys. Res.* 76:343-356.

Crozaz, G.; Haack, U.; Hair, M.; Hoyt, H.; Cardos, J.; Maurette, M.; Miyajima, M.; Seitz, M.; Sun, S.; Walker, R.; Wittels, M.; and Woolum, D. 1970. Solid state studies of the radiation history of the lunar surface. *Science* 167:563-566.

Dalgarno, A. and McElroy, M. B. 1970. Mars: Is nitrogen present? *Science* 170:167-168.

Dalgarno, A.; Degges, T. C.; and Stewart, A. I. 1970. Mariner 6: Origin of Mars ionized carbon dioxide ultraviolet spectrum. *Science* 167:1490-1491.

Gay, P.; Bancroft, G. M.; and Boun, M. G. 1970. Diffraction and Mossbauer studies of minerals from lunar soils and rocks. *Geochim. Cosmochim. Acta* 34, *Supplement* I, 481-497.

Gerstenkorn, H. 1955. Uber Gezeitenreibung beim Zweikorperproblem. *Z. Astrophys.* 36:245.

Gibert, J. and Oró, J. 1970. Gas chromatographic-mass spectrometric determination of potential contaminant hydrocarbons of moon samples. *J. Chromatog. Sci.* 8:295-296.

Gibert, J.; Flory, D.; and Oró, J. 1971. Identity of a common contaminant of Apollo 11 lunar fines and Apollo 12 York meshes. *Nature* 229:33-34.

Gold, T. 1969. A remarkable glazing phenomenon on the lunar samples. *Science* 165: 1345.

Goldreich, P. 1966. History of the lunar orbit. *Rev. Geophys.* 4:411.

Herr, K. C. and Pimental, G. 1969. Infrared absorption near three microns recorded over the polar cap of Mars. *Science* 166:496-499.

Holland, H. D. 1964. *On the chemical evolution of the terrestrial and cytherean atmospheres*, pp. 86-101. New York: Wiley.

Hubbard, J. S.; Hardy, J. P.; and Horowitz, N. H. 1971. Photocatalytic production of organic compounds from CO and H_2O in a simulated Martian atmosphere. *Proc. Nat. Acad. Sci.* 68:574-578.

Ingersoll, A. P. and Leovy, C. B. 1971. The atmospheres of Mars and Venus. *Ann. Rev. Astron. and Astrophys.* 9:(in press).

Kuiper, G. P. 1970. *The threshold of space,* ed. M. Zelikoff, London: Pergamon.

Lederberg, J. and Cowie, D. B. 1958. Moondust. *Science* 127:1473.

Leighton, R. B. and Murray, B. C. 1966. Behavior of carbon dioxide and other volatiles on Mars. *Science* 153:136-144.

Leighton, R. B.; Horowitz, N. H.; Murray, B. C.; Sharp, R. P.; Herriman, A. H.; Young, A. T.; Smith, B. A.; Davies, M. E.; and Leovy, C. B. 1969. Mariner 6 and 7 television pictures: Preliminary analysis. *Science* 166:49-67.

Leighton, R. B. and Murray, B. C. 1971. One year's processing and interpretation–An overview. *J. Geophys. Res.* 76:293-296.

Leovy, C. B. 1969. Mars: Theoretical aspects of meteorology. *Applied Optics* 8:1279-1286.

Leovy, C. B. and Mintz, Y. 1969. Numerical simulations of the atmospheric circulation and climate of Mars. *J. Atmos. Sci.* 26:1167-1190.

Leovy, C. B.; Smith, B. A.; Young, A. T.; and Leighton, R. B. 1971. Mariner Mars '69: Atmospheric results. *J. Geophys. Res.* 76:297-312.

Marcus, A. H. 1964. A stochiastic model of the formation and survival of lunar craters. 1. Distribution of clean craters. *Icarus* 3:460.

Margulis, L., ed. 1970. *Origins of Life,* Proceedings of the first conference. Gordon and Breach, N.Y.

Margulis, L., ed. 1971. Proceedings of the second conference on *Origins of Life:* Cosmic Evolution, Abundance and Distribution of Biologically Important Elements, Gordon and Breach, N.Y.

MacDonald, G. J. F. 1964. Tidal friction. *Rev. Geophys.* 2:467.

McConnell, J. G. and McElroy, M. D. 1970. Excitation processes for Martian dayglow. *J. Geophys. Res.* 75:7290-7293.

Moore, C. B.; Gibson, E. K.; Larimer, J. W.; Lewis, C. F.; and Nichiporuk, W. 1970. Total carbon and nitrogen abundances in Apollo 11 lunar samples and selected achondrites and basalts. *Geochim. Cosmochim. Acta* 34, *Supplement* I:1375-1382.

Morrison, G. H.; Gerard, J. T.; Kashuba, A. J.; Ganadharam, E. V.; Rothenberg, A. M.; Potter, N. M. and Miller, G. B. 1970. Elemental abundances of lunar soil and rocks. *Geochim. Cosmochim. Acta* 34, *Supplement* I:1383-1392.

Murray, B. C.; Soderblom, L. A.; Sharp, R. P.; and Cutts, J. A. 1971. The surface of Mars 1. Cratered terrains. *J. Geophys. Res.* 76:313-330.

O'Keefe, J. A. 1970. Apollo 11: Implications for the early history of the solar system. EOS 51:633. (*Transactions of American Geophysical Union*)

Opik, E. J. 1962. Surface properties of the moon. In *Progress in the astronautical sciences,* 215-260, ed. S. F. Singer. New York: Academic Press.

Opik, E. J. and Singer, S. F. 1960. Escape of gases from the moon. *J. Geophys. Res.* 65:3065.

Oró, J.; Gibert, J.; Updegrove, W. S.; McReynolds, J.; Ibañez, J.; Gil-Av, E.; Flory, D.; and Zlatkis, A. 1970. Gas chromatographic and mass spectrometric methods applied to the analysis of lunar samples from the Sea of Tranquility. *J. Chromatog. Sci.* 8:297-308.

Oró, J.; Updegrove, W. S.; Gibert, J.; McReynolds, J.; Gil-Av, E.; Ibañez, J.; Zlatkis, A.; Flory, D. A.; Levy, R. L.; and Wolf, C. 1970. Organogenic elements and compounds in surface samples from the Sea of Tranquility. *Science* 167:765-767.

Oró, J.; Updegrove, W. S.; Gibert, J.; McReynolds, J.; Gil-Av, E.; Ibañez, J.; Zlatkis, A.; Flory, D. A.; Levy, R. L.; and Wolf, C. J. 1970. Organogenic elements and compounds in Type C and D Lunar samples from Apollo 11. *Geochim. Cosmochim. Acta.* 34, *Supplement* I, 1901-1920.

Oyama, V. I.; Merek, E. L.; and Silverman, M. P. 1970. A search for viable organisms in a lunar sample. *Science* 167:773-775.

Plummer, W. T. and Carson, R. K. 1969. Mars: Is the surface colored by carbon suboxide? *Science* 166:1141.

Rasool, S. F.; Hogan, J. S.; Stewart, R. W.; and Russell, L. H. 1970. Temperature distributions in the lower atmosphere of Mars from Mariner 6 and 7 occultation data. *J. Atmos. Sci.* 27:841-843.

Rich, A. 1962. On the problem of evolution and biochemical information transfer. In *Horizons in biochemistry.* Kasha, M. and Pullman, B., eds., pp. 103-126. New York: Academic Press.

Ringwood, A. E. 1970. Petrogenesis of Apollo 11 basalts and implications for lunar origins. *J. Geophys. Res.* 75:6453-79.

Ross, H. P. 1968. A simplified mathematical model for lunar crater erosion. *J. Geophys. Res.* 73:1343.

Rubey, W. W. 1951. Geological history of sea water. *Bull. Geol. Soc. Am.* 62:1111. Reprinted in *The origin and evolution of atmospheres and oceans.* 1964. Brancazio, P. J. and Cameron, A. G. W., eds. Chap. 1, pp. 1-63. New York: Wiley.

__________. 1955. Development of the hydrosphere and atmosphere, with special reference to probable composition of the early atmosphere. *Geol. Soc. Am. Special Paper* 62:631-650.

Schorn, R. A.; Farmer, C. B.; and Little, S. J. 1969. High dispersion spectroscopic studies of Mars III. Preliminary results of 1968-69 water vapor studies. *Icarus* 11:283.

Schorn, R. A.; Spinrad, H.; Moore, R. C.; Smith, H. J.; and Giver, L. P. 1967. High dispersion spectroscopic observations of Mars II. The water vapor variations. *Astrophys. J.* 147:743.

Sharp, R. P.; Murray, B. C.; Leighton, R. B.; Soderblom, L. A.; and Cutts, J. A. 1971. The surface of Mars 4. South polar cap. *J. Geophys. Res.* 76: 357-368.

Sharp, R. P.; Soderblom, L. A.; Murray, B. C.; and Cutts, J. A. 1971. The surface of Mars 2. Uncratered terrains. *J. Geophys. Res.* 76:331-342.

Singer, S. F. 1968. Origin of the moon and geophysical consequences. *Geophys. J. R. Astron. Soc.* 15:205.

Singer, S. F. and Bandermann, L. W. 1970. Where was the moon formed? *Science* 170:438.

Socha, A. J.; Updegrove, W. S.; and Oró, J. 1970. Mass spectrometric analysis of lunar materials by probe technique. Presented at the Eighteenth Annual Conference on Mars Spectrometry and Allied Topics, San Francisco, Calif. p. B222.

Smith, R. N.; Young, D. A.; Smith, E. N.; and Carter, C. C. 1963. The structure and properties of carbon suboxide polymer. *Inorg. Chem.* 2:829.

Updegrove, W. S. and Oró, J. 1969. Analysis of the organic matter on the moon by gas chromatography-mass spectrometry: A feasibility study. In *Research in physics and chemistry,* ed. F. J. Malina. pp. 53-74. London: Pergamon Press.

Updegrove, W. S.; Oró, J.; and Lewis, R. 1970. Direct imaging mass spectrometric analysis of lunar materials by probe technique. Presented at the Eighteenth Annual Conference on Mass Spectrometry and Allied Topics, San Francisco, Calif., p. B219.

Urey, H. C. and MacDonald, G. J. F. 1970. Origin and history of the moon. In *Physics and astronomy of the moon,* ed. Z. Kopal. New York: Academic Press.

Wachi, F. M.; Gilmartin, D. E.; Oró, J.; and Updegrove, W. S. 1971. Differential thermal analysis and gas release studies of Apollo 11 samples. *Icarus,* 15:304-313.

Wachi, F. M.; Gilmartin, D. E.; Stukey, W. K.; Knight, V. L.; Oró, J.; and Updegrove, W. S. 1970. High-temperature high-resolution mass spectrometric studies of Apollo 11 lunar fines. Presented at the Eighteenth Annual Conference on Mass Spectrometry and Allied Topics, San Francisco, Calif., p. B264.

PARTICIPANT INDEX

SUBJECT INDEX

MIX
Papier aus verantwortungsvollen Quellen
Paper from responsible sources
FSC® C105338

Printed by Books on Demand, Germany